AF740323

Mixed Motives, Uncertain Outcomes

Mixed Motives, Uncertain Outcomes

Defense Conversion in China

edited by
Jörn Brömmelhörster
John Frankenstein

LYNNE RIENNER PUBLISHERS
BOULDER
LONDON

23 JAN 2003

PLC394
MIX

338.4762330951
mix

2007861

Published in the United States of America in 1997 by
Lynne Rienner Publishers, Inc.
1800 30th Street, Boulder, Colorado 80301

and in the United Kingdom by
Lynne Rienner Publishers, Inc.
3 Henrietta Street, Covent Garden, London WC2E 8LU

© 1997 by the Bonn International Center for Conversion. All rights reserved

Library of Congress Cataloging-in-Publication Data
Mixed motives, uncertain outcomes : defense conversion in China /
edited by Jörn Brömmelhörster and John Frankenstein.
Includes bibliographical references and index.
ISBN 1-55587-710-9 (hardcover : alk. paper)
1. Defense industries—China. 2. Military planning—China.
3. China—Defenses. I. Brömmelhörster, Jörn, 1961– .
I. Frankenstein, John.
HD9743.C623M59 1997
338.4'76233'0951—dc20 96-27539
CIP

British Cataloguing in Publication Data
A Cataloguing in Publication record for this book
is available from the British Library.

Printed and bound in the United States of America

The paper used in this publication meets the requirements of the American National Standard for Permanence of Paper for Printed Library Materials Z39.48-1984.

5 4 3 2 1

Contents

Part 4 The PLA

Part 5 Concluding Perspectives

Tables and Figures

Tables

Figures

Foreword

The changes wrought in the international security arena by the end of the Cold War constitute one of the principal stories of the late twentieth century. The economic and political reemergence of China, another top story, is one of the consequences of these changes. Diminished international tensions have also resulted in a global downsizing of military forces and consequent reductions in the need to maintain large military-industrial complexes. One use for these repositories of vast national treasure is called "defense conversion." The Chinese government has been sending mixed signals to the international community about its arms production and conversion strategy. Military expenditures have been substantially increased, after a period of drastic reductions during the early 1980s. At the same time, significant portions of the armed forces have been demobilized, and the government announced the success of conversion from military to civil production with considerable fanfare; it also has tried to attract foreign investment, often with the aid of international organizations, to help with the process.

For China, conversion not only is intended to cope with the consequences of reduced military production but also constitutes a policy for modernization of the Chinese defense industrial base through imports of modern technology. From a purely economic perspective, it has not always functioned as well as is officially claimed.

A full understanding and examination of defense conversion requires a multidisciplinary approach; no single perspective can capture the complexity of the process. This book presents a variety of views from an international group of economists, political scientists, industry insiders, and business executives on China's endeavors at defense conversion. It is worth noting that every one of the contributors—all Chinese speakers—has observed China's conversion attempts up close.

Readers will find that the contributors bring their own perspectives—disciplinary, professional, personal—to this collection; inevitably, there are differences in data selection and interpretation, and the editors have not made any attempt to impose a set of uniform conclusions. Nevertheless, readers may find that certain similarities emerge in the findings.

In Chapter 1, coeditor John Frankenstein offers an overview of China's defense conversion and concludes that Chinese claims of significant, early success in conversion are as much politically inspired as they are reflections of any concrete progress. In Chapter 2, Kay Möller highlights the paradoxical relationships among the objective international situation, Chinese domestic politics, and Chinese defense conversion efforts. Feng-Cheng Fu and Chi-Keung Li provide a macroeconomic analysis of conversion's contributions to the Chinese economy, in which they find enormous diversity and low productivity, in Chapter 3.

One must also examine conversion from the Chinese perspective, and Chai Benliang provides a detailed business-system analysis of the conversion effort in Chapter 4. In the fifth chapter, Ka Po Ng tracks conversion policy through the pages of specialized journals published in the PRC.

Taking more particular views, Yitzhak Shichor (Chapter 6) analyzes conversion in the aviation industry, Arthur S. Ding (Chapter 7) explores the situation in the Third Line industries, and Richard Latham (Chapter 8) discusses conversion as a business process.

Although the bulk of conversion occurs in civilian defense industries, the People's Liberation Army also participates in commercial activities. Tai Ming Cheung (Chapter 9) and Min Chen (Chapter 10) examine the fascinating story of the "entrepreneurial army."

In the final two chapters of the book, coeditor Jörn Brömmelhörster compares conversion efforts in China and Russia, and the two coeditors summarize the themes highlighted in the previous chapters.

The contributions in this book create a fascinating story of one particular approach to conversion. The Bonn International Center for Conversion, in researching conversion globally, strives to make the findings available to both China and non-China experts and to allow comparisons to conversion strategies in other parts of the world.

In addition to thanking the authors, I would also like to recognize Beth Partin, Alisa Federico, and the staff of Lynne Rienner Publishers for their editorial assistance during the preparation of this publication.

Herbert Wulf
Director
Bonn International Center for Conversion

PART 1

The Issues

1

China's Defense Industry Conversion: A Strategic Overview

John Frankenstein

Jun-min jiehe,	[Combine the military and civil,
Ping-zhan jiehe,	Combine peace and war,
Jun-pin youxian,	Give priority to military products,
Yi min yang jun.	Let the civil support the military.]

—"The 16-Character Slogan," Deng Xiaoping, 1982

The conversion of the People's Republic of China's (PRC's) vast military-industrial complex (MIC)—the nuclear, ordnance, aviation, space, electronics, and shipbuilding sectors under State Council direction, plus enterprises run by the People's Liberation Army (PLA)—to production for the civilian market has been one of the continuing themes of the regime's post-1978 economic reform efforts.[1] The conversion program gets attention at the highest levels: It is accounted for in the mechanisms of the central Five-Year Plans and is directed by a Three Commission Liaison Group for Defense Conversion, which brings together officials from the State Planning Commission (SPC), the State Science and Technology Commission (SSTC), the military's Commission on Science, Technology, and Industry for National Defense (COSTIND), and a State Council office that deals with problems of the "Third Line industries" (that part of the MIC located in remote areas of central and southwestern China).[2]

Great success has been claimed. According to a March 1994 article in the overseas edition of the Chinese Communist Party's newspaper *Renmin Ribao* (*People's Daily*), 77.4 percent of the gross output value of the MIC in 1993 was in civilian products, up from the 8.1 percent produced in 1978; 50 percent of the cameras and 65 percent of the motorcycles made in China are said to be produced by defense factories.[3] The electronics industry asserted that, by 1992, 97 percent of its production was in civilian products.[4] In 1994, the ordnance industry claimed that 90 percent of its industrial output in southwest China was for the civilian market—motorcycles, minicars, heavy-duty trucks, cameras, refrigerators and other white goods, optical and electrical instruments, and machinery for oil production.[5] And in early 1995, the official *China Daily* reported that 1994 sales of civilian products made by the ordnance sector increased 31 percent over 1993, to reach 18.5 billion yuan (about U.S.$2.2 billion); the forecast was for another 30 percent increase during the year.[6]

In a paper delivered at a conference on conversion held in Beijing in mid-1995, Zhang Weimin, a NORINCO vice president, gave an upbeat account of the group's development: 157 large and medium-sized factories, more than 30 research and development (R&D) institutes, 200 sales companies, and 60 subsidiaries trading in 100 countries; the group's joint ventures (JVs) produced 40 percent of the motorcycles sold on the domestic market and aimed to produce 450,000 minicars and 20,000 heavy trucks per year by the end of the century.[7]

Other sectors of the MIC make an even broader range of products for the civilian market. The aerospace industry for instance, claimed that more than 70 percent of its 1993 output value was in goods and services for the nonmilitary marketplace.[8] For a breakdown of that output, see Table 1.1.

Table 1.1 Chinese Aviation and Space Ministry Products for the Civilian Market

Geosynchronous and meteorological satellites
Satellite ground stations
U.S. FAA-certified aircraft and aircraft sections and parts
Jet engines and turbines
Industrial control computers
Four kinds of passenger buses
Eleven models of light trucks
Eight models of heavy trucks
Twelve models of motorcycles and motorbikes
Typewriters
Textile manufacturing equipment
Food packaging machines
Medical equipment and supplies (including X-ray machines, artificial joints, and ultrasonic kidney stone crushers)
Industrial robots, lathes, and five-axis milling machines
Color television sets
Refrigerators
Air conditioners
Fans
Vacuum cleaners
Teflon cookware
Portable radios and tape players
Watches and clocks
Drills and cutters
Automobile body jigs
Hovercraft
Windmills
Metal castings and metal casting services (including the cast bronze Buddha on Lantau Island in Hong Kong, reportedly the largest in the world)

Source: Derived from an untitled promotional catalog issued by the China Aero-Space Civil Products Corporation, Beijing, 1991.

These conversion claims suggest remarkable accomplishments. Given the economic and political imperatives—and extreme difficulties—of conversion in the West and the former Soviet Union, the Chinese experience is of more than passing interest. But less prominently advertised problems and strategies behind China's conversion efforts also warrant examination. China is the land of big numbers—numbers that, upon examination, are not confirmable or susceptible to anything but the crudest analysis. The chapters in this volume look at conversion in China from a number of perspectives. In this chapter I propose to outline the strategic themes.

Some Definitions

What do we mean by *conversion* and how do we define its success? And what are the international and domestic Chinese political, military, and economic contexts of conversion and the imperatives that arise from those changing situations?

Conversion is not a precise concept; it goes far beyond the commonplace that defines conversion as using defense production facilities to make civilian products. Questions of economics, public policy, and management practice must be taken into account.

In the conventional wisdom of economics, conversion has a positive economic effect. Simply put, military production has a lower multiplier effect than production for the civilian market. In addition, spending on military goods is considered by some accounting practices as expenditure on nonproductive capital. Thus, in a militarized economy, opportunity costs are high: Conversion of "unneeded" defense capacity, it is thought, will recoup those costs and, at the same time, increase the efficiency and wealth-generating capacity of the economy. But conversion also must be factored into the eternal national security question, how much is enough? In this sense, it is part of an optimization problem that must balance different demands—military, ecological, political, social—and diminishing resources.

Some also see the shift to production of civilian goods as contributing to peace and stability. But if conversion leads to more economic efficiency, it also can result in a reduction of military capabilities either by cutting into defense production capabilities through outright demobilization of facilities or, because of cuts in defense procurement, by forcing the defense sector to turn to the market for survival. The result, whether intentional or not, can be, in a phrase, "industrial disarmament."

Difficulties start to arise when we try to operationalize conversion. There are many definitions, including the following:

- *Production changeover:* The same equipment and lines that formerly made military items produce for the civilian market.

- *Diversification:* New facilities and lines for civilian production are created or acquired.
- *Redeployment and commercialization of military R&D and the exploitation of spin-offs:* For instance, work in remote-sensing technologies shifts from military intelligence to environmental surveillance; advanced aircraft radar is commercialized; or battery technologies used in smart weapons are applied to developing electric automobiles. Either defense firms themselves or, perhaps more importantly, defense research institutes that develop classified technologies and processes with potential civilian applications may be involved here.
- *Demobilization and retraining:* The efforts and talents of the people of the military establishment, both soldiers and workers, are redirected.
- *Surplus weapons management:* Difficult questions about recycling (tanks into tractors, or scrap), security, and sales must be considered.
- *Base closures:* This involves the disposal of real estate no longer required by the military and the tough question of managing the economic dislocation caused by base closures.
- *Re-engineering an entire economic sector:*[9] This dimension involves concentration (including divestment or closure) of redundant defense industrial capacity and may not precisely fit into "conversion." Nevertheless, it is an industrial response to diminished demand. It means the rationalization of the defense industrial base as a whole, from prime contractors down to second- and third-tier suppliers. It also signifies reshaping the industrial structure, from transactional hierarchies based on lowest-cost supply chains to tight, collaborative, partnership-based networks linking firms with distinctive competencies. In sum, it is "shrinking smart," in the phrase suggested by defense industry consultant Jerrold T. Lundquist.[10]

None of the assumed results of conversion is guaranteed by any of these approaches, which cannot be considered in isolation; conversion strategies must take all of them into account. Furthermore, the new resources and economic efficiencies achieved through conversion could result in a strengthened defense industrial base through reinvestment of profits and enhanced military capabilities. What we observe in the Chinese case, in fact, is an attempt to accomplish just that—conversion mixes these industrial strategies as part of a larger effort to modernize the Chinese military. Thus, as we shall see in the chapters in this book and as Paul Folta remarks in his earlier study of Chinese defense conversion, the overarching approach is "swords to plowshares . . . and better swords."[11]

Regardless of how we might define conversion, there are significant social questions and management issues that need to be addressed. What are the implications of conversion for employment? Will the highly trained and specialized defense industry engineers and technicians be able to make the shift or even be employable in an economy of conversion? How do communities dependent upon military support activities cope with reductions? And how does one reform the management and business strategy paradigms of defense businesses? Conversion is often treated as a relatively straightforward technical problem. But, in fact, it demands major shifts in corporate strategy and corporate culture to focus on markets and customers, away from practices that even in capitalist countries have been likened to those in a command economy. Fundamentally, conversion requires changes in ingrained work habits and mindsets.[12]

Four of the following chapters discuss these systemic realignments. Chai Benliang notes in Chapter 4 that although there has been considerable progress in corporatizing the MIC and in implementing enterprise reform and an "enterprise responsibility system," the stress on profits may affect military production negatively. Ka Po Ng looks at the process through the lens of official press coverage of Chinese defense conversion, and follows the discussion of organizational problems and solutions presented there (Chapter 5). If initially the defense sector moved into civilian production reluctantly, now the high-cost, low-profit demands of military production are disincentives. In Chapter 6 Yitzhak Shichor discusses conversion in the aviation sector and shows that it really denotes diversification, with no negative impact on military production capabilities. Richard Latham (Chapter 8) adds to the discussion by bringing his unique combination of China-watcher skills and a Western business perspective to bear on the issues of organizational transformation and business practice—he reminds us that conversion must be evaluated as a business proposition.

Finally, how can we measure success in conversion? Success here must mean more than pursuing make-work projects. Along with what some call "technical conversion," there must also be organizational and commercial conversion. Ultimately, successful conversion means earning a profit, modernizing organizations, and, indeed, being competitive in an increasingly global marketplace.

Conversion does not take place in a vacuum. It would not be such a topic of interest today without the recent major shifts in the international political and economic environment: the end of the Cold War, a refocusing of strategic concerns to regional issues, changes in military technology, and the increasing global dominance of the liberal economic paradigm. If we combine all these considerations, we can see that conversion is hardly a simple matter.

Chinese Contexts

Defense conversion in China is part of a reform process in both the economic and military spheres. Economically, China is in the midst of the marketization of a command economy, which requires a shift in the mechanism used for the allocation of most resources from a top-down, politically directed administrative system of plans and quotas to the marketplace. This process has encountered political resistance and, in the dislocations caused by the change, has given rise to numerous unintended negative consequences: a major increase in corruption throughout the country; problems with inflation at the consumer level (officially at about 25 percent in 1994, although it moderated in 1995 and 1996); questions over ownership, control, employment, and bankruptcy; dilemmas arising from 'bureaucratic entrepreneurialism"; and significant adjustment problems in dealing with the rules of the interdependent global system. Witness China's difficulty in coming to terms with entry conditions to the World Trade Organization (WTO) and the long-running dispute between China and the United States, China's largest export market, over intellectual property protection and market access.

But the reforms have had major successes as well. The economy has seen spectacular growth—about 10 percent annually since the reforms began. According to official statistics, Chinese nominal 1995 gross domestic product (GDP) is estimated to be about U.S.$695 billion, up from U.S.$370 billion in 1990, and is projected to continue its upward trend. China's total foreign trade volume reached almost U.S.$280 billion in 1995, making it the world's tenth most important trading nation. Foreign reserves in 1995 stand comfortably at around U.S.$70 billion. Despite the ups and downs of policy, China continues to attract substantial amounts of foreign direct investment—about U.S.$37 billion in 1995, seven times more than that at the beginning of the decade—and increasing amounts of portfolio investment as well (about U.S.$5.5 billion in bonds and equities in 1994). The effect on the standard and quality of life for millions of Chinese, especially those in urban areas, is incalculable.[13]

A Changing Security Environment

The economic reform program with all its benefits, however, could not have taken off if China's perception of its security environment had not changed. In the 1960s and 1970s, China faced two threats: the United States and the USSR. Ideological and political differences with the Soviet Union transformed it from China's socialist big brother into China's main adversary. The Soviet Union built up its military along its border with China. China's initial response during this period was "people's war under modern conditions"—a strategy, based on the PLA's peasant war experience, of allowing deep penetration and then cutting the invader off by flank attacks on supply lines. In the mid-1970s, the strategy focused on "all-out nuclear war"; Chinese planning considered the initial conditions of battle and foresaw fierce main force battles along a forward position. The force structure of the PLA reflected these strategies: a ground force of over 3 million, a small air force, an even smaller navy, and a separate nuclear weapons service arm—the 2nd Artillery. With 4 million people in the military, China had the largest standing military establishment in the world.

But the political requirements of China in the post–Cultural Revolution period of the late 1970s called for a shift in priorities toward economic development regardless of any possible external threat. On the ideological front, the final battle between capitalism and socialism was to be indefinitely postponed, and the threat of "social imperialism"—the code phrase for the USSR—was downplayed. The balance of the strategic triangle shifted, as the United States and China played their respective Chinese and American cards against the USSR. However, Kay Möller points out in Chapter 2 that even as the political dynamic changed and China began its most recent move toward conversion, security issues appeared to hold back conversion. These include the unsuccessful war with Vietnam, the enhanced political capital of the PLA after Tiananmen Square, offshore claims of national sovereignty (which raise worries in other Asian capitals about Beijing's regional political and military intentions), and the technological shock of the Gulf War.

In this newly defined political environment, Beijing could justify diverting resources away from the military toward civilian needs. Indeed, in the early 1980s there was a debate over allocations: Should the military be built up at the same time as the economy, or should the military wait until the economy was strong? In the final outcome, the military was put on hold; military doctrine now called for China to prepare for an "active defense" and limited (in time and space), local war. At the same time, the military started down a long road toward regularization and professionalism. Ranks, abolished during the Cultural Revolution, were reintroduced; military academies were strengthened and a National Defense University was opened in Beijing. In the mid-1980s there was a major reduction in force: Many older officers, veterans of the PLA's peasant army days, retired and parts of the army were transferred out of the PLA as members of a new People's Armed Police (PAP) force.

Organization

The industrial complex that is now undergoing at least partial conversion and that had grown to service China's military always occupied a significant fraction of China's industrial infrastructure. Forty-one of the 156 "key projects" that were built with Soviet assistance in the 1950s were, in the words of Stanford scholars John Lewis and Xue Litai, "directly connected to the output of conventional weapons." They cite a former Soviet adviser as saying that "most" of the credits extended to China by the USSR in the 1950s went to defense—"war equipment, machinery for munitions plants and other . . . military enterprises"; some "thirty percent of the roughly 250 industrial facilities eventually outfitted by the Soviet Union were in the defense sector."[14] Following the Soviet model of a national security state, the MIC was a dominant force in the industrial sector of the economy.

The structure of the MIC has been fluid. Today it is composed of two distinct parts: the ministries and corporations under the State Council—including enterprises under local provincial government—on the one hand, and PLA organizations on the other. It is important to understand the distinction; the former are the "defense industries," whereas PLA factories, in the military chain of command, are the "military industries." The civilian-run defense industries evolved from a collection of secret, numbered, "machine-building" industrial ministries headed by military officers to an array of civilian-run, profit-seeking corporations that have increasingly sought to diversify their activities. PLA economic organizations started as units devoted to internal, self-supporting supply, distribution, and military logistics but have expanded their operations to the civilian sector; these include operations run from the top command organizations down to the seven military regions, the twenty-four group armies, and subordinate local units. Tables 1.2 and 1.3 show first the evolution of the civilian-run defense industries and second the current organization of the MIC.[15]

The MIC is geographically dispersed: In the 1960s and early 1970s, a large part of the MIC was constructed in remote areas of southern and western China; according to Chinese officials, 55 percent of China's defense factories are in this "Third Line."[16] Intended to be a strategic relocation, in fact this move was a huge expense and is now deemed irrational; in private conversations, Chinese defense industry analysts acknowledge that the Third Line movement—like Mao's other mass campaigns—led to great waste and, in retrospect, was a mistake based on an inappropriate reading of the Soviet experience. Estimates are that over 50 percent of Chinese national investment during that period went to the Front, but infrastructure and communications—roads, rail lines, tunnels—in these remote and mountainous areas absorbed 80 percent of that investment. The Third Line's isolation created local economic imbalance and dependence on military orders—between one-third and one-half of total industrial output in Sichuan, China's most populous province, depends on the MIC.[17] Today, these Third Line industries are being encouraged to make a "triple jump"—"jump out of the backwater, skip to coastal areas, pole vault overseas"—and by 1992 it was claimed that they had established eight hundred enterprises along the coast and in cities and special economic zones. To do this the state is spending over U.S.$1 billion; it may take well into the twenty-first century to complete the process.[18] The remote locations and decentralized organization of this major portion of China's defense industry have had a major impact on

Table 1.2 Evolution of the Defense Industries

MBI	Ministry/Corp (1982)	Ministry/Corp (1988)	Ministry/Corp (1993)
2	Nuclear Energy Ministry	Ministry of Energy Resources (MER)	China National Nuclear Corp. (MER broken into Coal and Electric Power Ministries)
3	Aviation	Ministry of Aerospace (MAS) (combines Aviation and Space Industry Ministries)	Aviation Industries of China (AVIC) (MAS broken into AVIC and China Aerospace Corp [CAsC])
4	Electronics Ministry	Machine Building and Electronics Industry Ministry (MMBEI)	Ministry of Electronics Industry[a] (MEI) (MMBEI broken into MEI and Ministry of Machine Industry, a reversion to the first MBI, or civilian, MBI organization)
5	Ordnance Ministry	MMBEI	China North Industries Corp. (NORINCO [Gp]) (ex-Ordnance Ministry, MMBEI)
6	Ship Construction Corp.	No change	No change
7	Space Industry	MAS	CAsC
8	Missiles (merged with space industry in 1981)		

The information in this table comes from private conversations with knowledgeable foreign observers in Beijing. On the consolidation of enterprise groups, see *Jiefangjun Bao* [Liberation Army Daily], 18 February 1995, cited in *Defense Science, Technology and Industry Digest (DSTI),* Hong Kong, February 1995.

Note: a. Given the trends toward corporatization, it seems paradoxical that electronics, one of the most successful of the "converting" sectors, remains under a ministry. According to industry sources, most electronics factories have been placed under provincial and local control or had consolidated into enterprise groups, and thus there was little, if anything, for the center to corporatize. At the same time, MEI has responsibilities in the development of telecommunications and technology acquisition, and thus remains. We would add, however, that corporatization has not been smooth—insiders at AVIC, for instance, say that the organization still behaves less like a profit-seeking corporation and more like a bureaucratic ministry. Indeed, they say, there is a problem of identity—one day they're told to go out and make money, and the next they're told to pay attention to political objectives.

the industry's effectiveness in both military production and conversion. In Chapter 7 Arthur S. Ding, working from official records and his own on-site investigations, explores the issues facing development of the Third Line: outflow of manpower and capital and the increasing economic disparity between coastal and inland regions. Paradoxically, Ding finds, the policies aimed at helping the Third Line have, in fact, harmed it.

Table 1.3 Current Structure of China's Military-Industrial-Commercial Complex

Central Military Commission (CMC)			
General Staff Dept. (GSD)a	General Political Dept. (GPD)	General Logistics Dept. (GLD)a	People's Armed Police (PAP)b
Poly Group Corp. (CITIC connection: arms, technology trade)	**Kaili (Carrie) Corp.** (communications equipment, publications)	**Xinxing Corp.** (clothing, food, construction materials)	**Jingan Equipment ImpEx Corp.** (police/security equipment, small arms, ammo)
China Huitong Corp. (Telecom)	**China Tiancheng Corp.**	**999 Enterprise Group** (pharmaceuticals)	**Anhua Development Corp.**
Second Artillery (Strategic Forces) Shanhaidan Co.		**San Ding Trading**	
Air Force China Lantian Corp. United Airlines			
Navy China Songhai Industrial and Commercial Corp.			
Bureau of Military Equipment and Technology Corp. (BOMETEC) Ping He Electronics (connections with Everbright)			
Communications Dept. China Electronic Systems Zhihua Corp (connections with Ministry of Foreign Trade and Economic Cooperation)			

There are also numerous PLA enterprises under the seven Military Regions, twenty-four Group Armies, and local authorities (the special Economic Zones), including:

Beijing MR: Huabei Jinghai Enterprise Corp.
Shenyang MR: Dongbei Jincheng Industrial Corp., Songliao Enterprise Group
Nanjing MR: Huadong Industrial Corp.
Jinan MR: Shandong Dongyue Corp.
Chengdu MR: Southwest Great Wall Development Corp.
Lanzhou MR: Northwest Industrial and Trading Corp.
Guangdong MR: Nanfang Industrial, Cuicum Enterprise United Group

This table indicates rough relationships within the bifurcated Chinese military-industrial complex. The Central Military Commission has primary authority over the military enterprises on the left; the State Council has primary jurisdiction over the ministries and corporations on the right. CITIC, the SPC and SSTC also provide input at the State Council level. COSTIND provides guidance and coordination to the ministries and corporations and acts as a bridge between the military and civilian authorities, although its authority may be waning as the CMC increasingly diversifies and converted defense plants come under provincial control. "Defense conversion" in China has many meanings, including the diversification of old State Council/ministerial defense firms into civilian production and the use of PLA assets to provide

(Notes continue on page 14)

Table 1.3 (continued)

State Council[c]			
Commission on Science Technology, and Industry for National Defense (COSTIND)[d]	Ministry of Machine Building[e]	Ministry of Electronic Industry	Ministry-Level Corporations
Xinshidai Corp. (exchanges, publications planning, marketing)		**China National Electronics Impex** (electro-optics, radar, crypto, EW)	**Aviation Industries of China** (AVIC) (ex–Aerospace Ministry)
Xiaofeng Technology and Equipment Corp. (computers, other high-tech)			**China Aero-Technology Impex** (CATIC)
Yuanwang Group plus at least 100 other companies, such as Shenzhen Heli S&T Development, China Galaxy			**China Aerospace Corp.** (CAsC) (ex–Aerospace Ministry)
Conversion Enterprises "Defense conversion" has led to the creation of new companies and enterprise groups, including "window" operations on the coast, joint ventures (such as NORINCO's motorcycle factories and Hua Mei Telecom), industry-based groups (such as the Huanghe and Zhenhua electronics groups), regionally centered groups (such as Guizhou's conglomerates), and new cross-sectoral groups, such as the COSTIND-sponsored Hainan Heping Corp., which draws on investments from the aviation, space, shipbuilding, and nuclear sectors. Defense conversion efforts are directed by COSTIND, the SPC, and the SSTC, aided by two COSTIND organizations, the Chinese Association for Peaceful Use of Military-Industrial Technology (CAPUMIT), and the China Defense Science and Technology Information Center.	**NUCLEAR INDUSTRIES** **China National Nuclear Corp.** (CNNC) China National Nuclear Instrumentation and Equipment Corp. (CNIEC) China Nuclear Energy Industry Corp. (NBC protection equipment) China Nuclear Equipment and Materials Corp. Rainbow Development Corp. (nuclear power)		**China Precision Machinery Impex** (optics, missiles) **Great Wall** (space launch) Beijing Wan Yuan Industry Corp. (space services) Chinese Academy of Space Technology (satellites) Chang Feng S&T Ind. Group[f] **China State Shipbuilding Corp.** **China North Industries Group (NORINCO [G])** (ex–Ordnance Ministry: arms, armor) China General Industrial Material and Equipment Supply Corp. China Yanxing National Corp. China North Optics and Electric Corp. **NORINCO** (Impex)

(*Table 1.3 continues*)

Notes to Table 1.3 (continued)

goods and services for the larger economy. In the former case, new enterprises may be created; in the latter, the PLA maintains ownership and control. However, as with any chart, not all connections can be shown. Informal relationships based on family, personal, or factional ties may often transcend the formal chain of command. Furthermore, the chart does not attempt to revolve to the enterprise level (for instance, it does not name the many enterprises that fall under NORINCO [G]); neither enterprise-level linkages between the defense and civilian sectors nor links with organizations and enterprises outside of China are shown.

Sources: Data on PLA/MR companies come from Chapter 9 of this book. See also H. Wulf, ed., *Arms Industry Limited* (Oxford: Oxford University Press), pp. 276–277.

Notes: a. The GSD & GLD oversee and coordinate with COSTIND external military sales.

b. Joint Public Security Bureau command.

c. Input from China International Trust and Investment Corp. (CITIC), State Planning Commission (SPC), State Science and Technology Commission (SSTC).

d. Control by CMC, guidance and coordination by State Council.

e. Ex-1 MBI, not much defense-related activity.

f. Chang Feng S&T Industry Group is typical of new forms of defense industry organization. It apparently is a kind of holding company, reporting to the 2nd Aerospace Academy and collecting 71 national research organizations under its wing. The enterprises under its direction belong to the space, aviation, ordinance, machine-building, and electronics industries and are spread out over 14 provinces. The product range will be broad: numerically controlled machine tools, medical and pharmaceutical manufacturing equipment, satellite and radar communications equipment, industrial process control systems, and 'special components.' (Source: China Defense Industry, 1 November 1994 in *DSTI,* November 1994.)

The Numbers Game

Official numbers vary as to the size and scope of the MIC. According to the *Encyclopedia of the Chinese Economy,* the State Council sector of the MIC today encompasses about 1,000 enterprises, each consisting of multiple factories, marketing organizations, and research units; more than 200 major research institutes (RIs); more than 300,000 engineers and technicians; and some 3 million workers.[19] These generally accepted figures may, in fact, be on the low side: Ka Po Ng cites China Association for Peaceful Use of Military Industrial Technology (CAPUMIT) official Jin Zhude to the effect that defense conversion "stabilized" the situation of 12 million people. Given the proliferation of operations encouraged by the reforms, shifting ownership and control patterns, and the deliberate vagueness of Chinese statistics, it is difficult to be more precise than these general figures.[20]

Individual sectors are quite large and diverse. For instance, under the general rubric of the aviation industry we find 6 universities and colleges, over 30 research institutes, and more than 200 trading companies and enterprises employing more than 500,000 workers, of whom 200,000 are engineers or qualified technicians. Enterprises under such an organization are not small; they are vertically integrated conglomerates. For example, the Chengdu Aircraft Engine Corporation, a "backbone" enterprise of the industry, employs 20,000 engineers and workers in 16 factories, 4 research institutes, 11 "branch companies," more than 40 joint ventures with outside capital, and 6 "window" enterprises—operations set up in the open coastal cities (and perhaps even overseas).[21] Defense industries in Shaanxi province, according to a 1993 report, include more than 130 enterprises and research institutes employing 280,000 staff and 90,000 scientists, engineers, and technicians—only Sichuan has a larger defense labor force.[22]

The Contradictions of Conversion

With cuts in military procurement arising from force restructuring and the general economic shift toward market forces during the 1980s, plus a decline in arms sales due to the end of the Iran-Iraq War (in which China supplied both sides), large parts of the Chinese military-industrial complex became redundant. Although reliable data on Chinese weapons production are not available in the open literature, we can get a sense of the decline when we consider that, according to the Stockholm International Peace Research Institute (SIPRI), Chinese exports of major weapons systems dropped from U.S.$3.1 billion in 1987 and U.S.$2.2 billion in 1988 to about U.S.$1.1–1.2 billion per year in the 1990s.[23] Under those circumstances, there would be a need to find a use for and to preserve those assets. Conversion would be the answer. However, another contradictory dynamic figures in as well—the continuing need for advanced military technology and the pressing problem of how to pay for its development.

That need was brought home to the leadership at three separate times since 1978. First, during the 1979 Sino-Vietnamese border war, the lack of intelligence, poor maps, and inadequate communications led to serious Chinese losses. Second, the high-tech weapons used during the Gulf War clearly showed the Chinese military leadership how far they were behind in their own armaments. And third, in the more immediate tactical context, the 1992 decision to enforce Chinese sovereignty over claims in the South China and Yellow Seas has become an impetus for improving the Chinese navy and for developing force extension capabilities; to do so the military has acquired Soviet fighter-bombers and Kilo-class submarines and has increased R&D into long-range communications, over-the-horizon radar, and air-to-air refueling. Bluster over Taiwan, inspired by the island's democratic progress, must only reinforce concerns over China's real military capabilities. To truly improve the military would require the redirection, if not expansion, of expenditures to develop new force structures and technologies.

At the same time, the economic reforms and the industrial responsibility system that applied to the economy as a whole have had a major effect on the structure and practices of the protected military-industrial complex. Managers and cadres now must deal with market forces, contracts, and taxes instead of the more familiar system of plans, allocations, and quotas. The new system led to behavior that could have negative consequences even for the PLA; for instance, according to a study of the Chinese air force (PLAAF) written by two former U.S. defense attachés, the export models of the F-7 (a modified version of the MiG21 that contained advanced Western avionics) were not purchased by the PLAAF because the Ministry of Aerospace Industries required payment in hard currency.[24] Indeed, the very structure of the system changed. In 1993 the Aerospace Ministry split into the China Aviation and China Space Corporations (eleven years earlier, China's aviation and space sectors had been separate ministries and, according to officials of the Space Corporation, the bureaucratic split remained despite the merger) and other ministries were shuffled and renamed. Bureaucratic entrepreneurialism has flourished, as state and party organizations—spurred by the requirements to be self-supporting—have taken advantage of reform's loosened rules and devolved authority.

How has China dealt with the seeming contradiction between cuts in procurement and growing commercialism on the one hand and the demand for new military technologies and capabilities on the other? Although the strategy has not been made explicit, it would appear that the People's Republic of China has opted for a "pockets of excellence" approach. The PLA has created "fist" or "rapid reaction units"—elite, modernized brigades scattered among the twenty-four group armies. With regard to technology, there seems to have been particular attention paid to the development of mobile ballistic missiles; those fired into international waters north of Taiwan in mid-1995 all appeared to have found the target zone. But these remain "pockets"—the larger force has yet to modernize, and the forces of economic reform have offered new, if commercial, targets for the PLA.

The PLA: A Special Case

One of the most dramatic examples of the new commercial spirit has come in the extension and expansion of PLA activities into the commercial domestic and international market. These striking examples of diversification, although they do not really fit a strict definition of conversion, are worth elucidating. Although both military leaders and intellectuals have repeatedly called for the PLA to get out of commercial operations—because of concern over corruption and fear that a concentration on moneymaking will sap the military ethic and further spread corruption through the Chinese system—there are few signs that the PLA is about to retreat on the commercial front. The large corporations run by the top command and the many local operations under military region and group army control look as if they will be around for a while.[25]

For example, MIC organizations and the PLA have established special economic and development zones in Guangdong. Jingan, the commercial branch of the People's Armed Police, has marketing operations for ammunition and small arms in the United States. Units connected with China's strategic missile service arm run ice-cream parlors in Beijing. Companies affiliated with Poly Group are active on the Hong Kong market and have formed groups making investments throughout "Greater China"—which includes, of course, Taiwan.

Some of these operations are quite substantial. The General Logistics Department (GLD) runs two of the largest: the Xinxing and the 999 Corporations. Xinxing was set up in 1984 with the explicit aim of becoming "one of China's top business giants."[26] It was incorporated as the Xinxing Group in 1989, and describes itself as having "close to 100 enterprises . . . in 20 provinces and municipalities." Its range of businesses includes import-export, marketing, industrial development, minerals and chemicals, real estate development, tourism, and advertising. It controls at least fifty factories employing 140,000 in the manufacture of textiles, leather shoes, machinery, electrical appliances, iron and steel, automobiles, medicine, and ships and has offices in Hong Kong and elsewhere.[27] The 999 Corporation is primarily involved in pharmaceuticals; operates thirty-four enterprises; has branches in Germany, the United States, Russia, Thailand and elsewhere; and has expanded its operations to include real estate development, electronics, food, clothing, and securities and trust investment services.[28] Large "999" advertising displays appear worldwide: Not only do they decorate the flight tower at the Capital Airport in Beijing, but they also add a distinctive touch to Hong Kong's spectacular harborside, nighttime, neon glow and can even be found amid the billboard clutter of Times Square in Manhattan. To be sure, the PLA has always been involved in economic activities, mostly to guarantee self-sufficiency and promote civic action—legacies of its origins as a peasant guerrilla force. Although we might wonder at an "entrepreneurial army," what the PLA does only reflects what is happening throughout China.

From Bureaucratic Entrepreneurialism to Foreign Investment

Other organizations throughout the system have also exploited the new situation: Universities, research institutes, and even State Council economic organs have started commercial consulting companies, and organizations rent out their facilities to private enterprise. The Beijing University of Aeronautics and Astronautics runs over twenty enterprises that have undertaken commercial R&D on projects ranging from ultralight aircraft to erasable optical disks and oil burners for industrial boilers and furnaces.[29]

In this flurry of diversification, foreign investment also plays a part. The Shanghai joint venture between McDonnell-Douglas and the aviation ministry, which has produced U.S. Federal Aviation Administration (FAA)–certified MD-80 passenger jets, is one of the best-known joint ventures. For the MIC, joint ventures are attractive for several reasons: They facilitate the shift to a market economy and bring needed capital, market knowledge, and management skills. At the same time, they may also bring advanced technology.[30] Table 1.4 suggests the range of civilian operations and technologies the MIC is seeking from abroad.

Table 1.4 Some MIC Joint Venture Proposals

Organization	Location	Joint Venture Sought
China Institute for Radiation Protection	Taiyuan, Shanxi	Medical uses of radiation, radiation-measuring instruments
Huanguang Industry Corporation	Hanzhong, Shaanxi	Civil engineering projects, cosmetics
Guizhou Aircraft Industrial Corporation	Anshun, Guizhou (with offices in Beijing, Shanghai)	Minicars, food processing equipment, household electrical appliances
Shenyang Aero-Engine Research Institute	Shenyang, Liaoning	Engineering design services, textiles
Qing'an Space Equipment Corp	Xi'an, Shaanxi	Air conditioner compressors
Jiangnan Machinery Plant	Xiangtan, Hunan	Industrial sewing machines, electronics, electrical appliances, chemicals
Sanjiang Machinery Plant	Yibing, Sichuan	Textile machines, heavy-duty truck parts
China South Aero-Engine and Machinery Corporation	Zhuzhou, Hunan	Optics, turbine engine parts, multilayer printed circuit boards

Source: Jin Zhude, ed., *Guide to International Cooperation and Investment with Enterprises of China's Defense Industry* (n.p.: China Association for Peaceful Use of Military-Industrial Technology, 1993).

And, of course, the MIC taps into international markets on its own. Not only have subsidiaries of NORINCO (Group) placed shares in the Shenzhen and Shanghai stockmarkets, but also the larger group—with offices around the world, including Dubai and Moscow—floated a U.S.$165 million closed-end fund on the Singapore and Dublin exchanges.

Assessing the Outcome

Certainly the claims made and the evident diversity and scope of activities attributed to conversion are impressive. There is no doubt that the MIC can produce goods and services for the civilian market, and it is evident that the sector's managerial and financial sophistication is increasing. But we must go beyond the obvious impact of the reform program on the MIC and the resulting tactics of diversification. For a fuller picture, we should ask what goals are being met by all this effort and whether problems are being encountered. We must also return to the problematic nature of Chinese statistics—it may be that defense conversion represents some large percentage of industrial output value, but gross output value statistics are full of double counting and do not take China's high inflation rates into account. We do not have the real numbers from which those percentages are calculated, nor do we really know how these numbers relate to economic statistics (GDP, exports, etc.). The numbers may even obscure a declining situation if we consider the poor condition of state-owned enterprises (SOEs). Indeed, if we accept that a high percentage of defense industry output is for the civilian market and take at face value the claims of a number of enterprises that they dominate their particular conversion sectors, we can only conclude that the vast majority of defense plants account for a very small proportion of any kind of output, civilian or military.[31] In other words, even with the best of intentions, it is difficult to come to a realistic assessment of the effort, and when we come to examine the strategies driving conversion in China, we are forced to raise some questions about the real intentions of the conversion effort.

Intentions

If we go beyond the achievement claims of the MIC, we can see that conversion in the minds of many Chinese officials means not only enhancing the civilian economy but also aiding Chinese military production capabilities.[32] The "16-Character Slogan," attributed to Deng Xiaoping and first articulated as early as 1978, spells it out: "Combine the military and civil, combine peace and war, give priority to military products, let the civil support the military." To be sure, the slogan is ambiguous and could be interpreted to mean either (or both) a short-term strategy to use defense conversion proceeds for immediate defense modernization or a long-term strategy, argued in crucial debates of the early 1980s, to develop the civilian economy before investing in military modernization (see Box 1.1). Still, even as high officials such as State Planning Commission chief Zou Jiahua have repeatedly called for continuing reform, they also stress the need to give priority to military production.[33] Thus, one of the goals of the 999 Corporation is to "play a positive role in developing the Army's production and in making up for the inadequacies in military spending."[34]

The same intention has been expressed lower down on the Chinese bureaucratic ladder. For instance, speaking at a defense conversion

Box 1.1 The "16-Character Slogan"—The Chinese View*

Jun-min jiehe,	Combine the military and civil,
Ping-zhan jiehe,	Combine peace and war,
Jun-pin youxian,	Give priority to military products,
Yi min yang jun.	Let the civil support the military.

Like many of the utterances of China's leaders, Deng's "16-Character Slogan" is delphic. Authoritative Chinese commentators interpret it through the lenses of Marxism–Leninism–Mao Zedong Thought. What follows below is a summary of the discussion of the slogan taken from the "Chapter on Theory" by Chen Dedi and Bao Qubing, two members of the book's editorial committee, in *Chinese Defense Conversion.*

The authors define defense conversion as "the use of military-industrial production capacity and military-industrial technology to serve the national economy" (29/17). But they take a long-term view typical of Marxist thought: "As social development advances to higher levels, peace and development become the demands of human society, the class struggle abates and . . . although national defense industries will not completely disappear, they are still necessary in a small scale. . . . in this context we say defense conversion is the product of a certain stage of social development" (30/17). Indeed, the argument runs, the very existence of military industries provides a country a certain measure of security, and, in any event, modern warfare requires high technology equipment—if these "are not developed and stored in peacetime, once war breaks out, there is no time to obtain them" (36/19). Against this background, the authors claim that the "16-character program (*fangzhen*) is the military-industrial construction policy (*fangzhen*) of China and is also the policy (*fangzhen*) for the development of the national economy" (42/23); it also represents a "leap (*yici feiyue*) in military industrial economic theory" (35/19).

The authors dissect the sentences of the slogan (somewhat out of order) as follows:

- *Combine peace and war.* "In the development of military industries consideration must be given to wartime supply . . . effective peacetime development and reserve capacities for military production" (38/20).
- *Combine the military and civil.* Here, the authors urge the development of production technologies that can produce both civilian and military goods and suggest (mistakenly in the view of this writer) that "the technology for the manufacture of military equipment and that for the manufacture of civilian products are mutually interchangeable." In their words, there is "duplication and compatibility" between military and civilian requirements (38/21).
- *Give priority to military products.* This refers to the "business activities" of military industries: "They must first of all satisfy the requirements of the state for military goods." At the same time, the state "should ensure that military industries are able to achieve the national security requirements" by providing "macroeconomic guarantees" of profits. If military industries make high-profit civilian

Box 1.1 (continued)

products that get in the way of military production, then it must "sacrifice" those profits to meet military needs (39–40/21–22).

- *Let the civil support the military.* "In the narrow sense," this means using the profits from civilian products made with military production technology to upgrade that technology and those products. In the 1980s, the authors say, the money earned by some factories "developing civilian products was used to develop military products." In a more general way, defense expenditures come from tax revenues, and thus this part of the slogan also refers to defense allocations from the (expanding, we hope) national tax base (40–41/22).

There is certainly nothing in the text that suggests the Chinese authorities see conversion as anything but a mode of strengthening their national security position. "For thousands of years there have been periods of peace followed by periods of war," they write (35/19). Conversion in peacetime, in theory at least, will strengthen the defense industrial base for the wars to come.

Source: This box draws upon Chen Dedi and Bao Qubing, "Chapter on Theory," in Cao Shixin, ed., *Zhongguo Junzhuanmin* [Chinese Defense Conversion] (Beijing: Zhongguo Jingji Chubanshe, 1994), pp. 27–49 and the *FBIS China Daily Report Supplement,* 27 July 1995 (*FBIS-CHI*-95-144-S) which provides a translation of the book. In some cases the translation used here differs slightly from that provided by *FBIS.* Page references in the text refer first to the original, second to the *FBIS* text. It is not clear why the authors reverse the order of the first two elements of the slogan in their analysis.

conference held in Beijing in 1991, a senior engineer in the aerospace industry said: "Among the dual tasks of serving national defense construction and serving economic construction, the national defense construction goes first and the economic construction second. . . . the system for both military and civil uses is a long-term stable military system." A Hubei provincial official said that the "guiding principle" for conversion was to "treat [military and civilian] equally without discrimination, [but to] put military production first, and give it appropriate preferential treatment" (by which he meant tax breaks and first call on infrastructure).[35]

Dual Use: Spin-off and Spin-on

Others have promoted dual-use technologies in the conversion effort. Indeed, some commentary in 1995 suggests that for some conversion means "spin-on" as much as it does "spin-off." No less than Central Military

Commission (CMC) Vice Chairman General Liu Huaqing was reported by *Liberation Army Daily* to argue at a January 1995 national conference on cooperation and coordination work in the military industries that "we must seize the opportune time of the end of the Eighth Five-Year Plan [1995] and the Ninth Five-Year Plan to push our national defense science and technology and weaponry onto a new state." According to the paper, he said that contributions from civilian industry, the Chinese Academy of Sciences, and the university system are "component parts" and the "foundation for developing science and technology industries for national defense." China, he said, "should pay attention to turning advanced technology for civilian use into technology for military use."[36] A Xinhua news-service commentary, also dating from January 1995, noted a number of "spin-on" developments: civilian industry "solved a large number of sophisticated technology problems crucial to the production of nuclear weapons, nuclear submarines, guided missiles and satellites [and] new materials."[37] And Huai Guomo, the COSTIND deputy director in charge of conversion, has pointed out: "The trend toward the interchangeability of military and civilian technology is increasing, and this provides a solid technological basis for the rapid modernization of national defense and the constant upgrading of weaponry."[38]

Arguing that dual use makes a contribution to the civilian economy while keeping military production lines running, an official of the Chinese Defense Science and Technology Information Center pointed out that in times of crisis this capability can strengthen deterrence; remarking on foreign experiences with dual use, he said, "the development of dual-use technologies provides the opportunity of developing military technologies in disguised form . . . [and] can save the costs in weapons systems and [military] modernization."[39] A Chinese scientist, who formerly designed intercontinental ballistic missile (ICBM) engine systems, told me privately that money from civilian satellite launches goes toward the salaries of scientists and engineers working on the military side of the Chinese missile program. Significantly, Chinese defense conversion officials have said that conversion had "fruitful material results": The economic efficiency of the military industries improved, and a "new generation" of military-industrial enterprises was created.[40] In other words, the *intention* of conversion appears to be not only to utilize redundant facilities but also to apply the proceeds of conversion to the maintenance of the defense industrial base and to further military modernization. If demilitarization is one objective of conversion, we do not find it in the Chinese case. Rather, we find here a strategy of rationalization and diversification.

Results

It is one thing to articulate a strategy and another thing to implement it, and in implementation, competing issues can emerge. When CAPUMIT official

Jin Zhude was asked about results by a newspaper reporter, he said that a little profit was being made overall, and "the remarkable thing" was that with conversion, "unemployment was avoided."[41] Jin had earlier written that "timely conversion" had many benefits: It "avoided unemployment, close-downs of factories, and social instability caused by sharp reduction of military orders."[42] In light of Chinese anxieties over stability, these concerns must be one of the key factors motivating conversion—if not the major one.[43]

It is also clear that the effort has its problems. To begin with, the MIC is part of the state sector, which is in serious decline and now accounts for less than 50 percent of Chinese industrial output. If in the early 1990s the conventional wisdom was that about one-third of SOEs lost money and another quarter just broke even, by late 1995 some Chinese economists were saying that 70 percent of SOEs were "in the red."[44] Zhu Rongji is reported to have said that the majority of the SOEs in trouble are from the defense sector; their plant capacity utilization rates are between 10 and 30 percent.[45] A Chinese media report termed 1994 a "good year" for the ordnance industry: Only 38 percent of China's ordnance enterprises were unprofitable, an improvement over 1993—when 50 percent were loss makers—and "the worst year" (unidentified) when 70 percent lost money.[46]

If we examine the record since 1979, we find an interesting dialogue in progress: Beijing calls for greater efforts in conversion and, a few months later, industry responds with assertions of progress. Because this has been going on for well over a decade, we can only conclude that conversion—as in other countries—has met with resistance from industry. In part this could arise from the foreign currency that flowed into defense industry accounts from foreign military sales. It could also be that the central government's support for conversion has been considerably less than generous. According to an OECD Development Center report on Chinese defense conversion, the state has allocated about 20 billion yuan to conversion since 1979; of this, 11 billion yuan came from the central government and 9 billion from the provinces.[47] However, this total represents only about 2 percent of the reported amount allocated to the state sector for "technical upgrading" since 1978.[48] In fact, according to the OECD report, much of the cost of conversion is being covered by bank loans—which, in the case of the larger and economically weak state-owned sector of which the MIC is a part, are as much bailouts as they are commercially viable debt.[49] Thus, despite Beijing's urgings, conversion appears to receive paltry direct financial support—that is, its share of the support offered by Beijing seems to be proportional to its contribution to the overall economy. To be sure, the government's resources are constrained, reform has pushed managerial responsibility down into the system, and much of the converting defense sector has been placed under provincial and local authority. One result has been, paradoxically, increased participation in conversion, even if *economic success*—profits—does not always follow. By the early 1990s, according to *Renmin Ribao,* only 40 percent of the MIC's enterprises were involved in civilian production; five years later, the OECD report indicated that about 80 percent of the MIC was somehow involved.[50] Furthermore, if one looks at the defense conversion products offered in defense conversion exhibitions or to foreign business people for joint venture investment, one is struck by their enormous range: from primary and intermediate industrial goods to finished products. The extremely high degree of vertical integration in the sector—and the inefficiencies resulting from lack of scale—that this suggests is remarkable.

Feng-Cheng Fu and Chi-Keung Li, using macroeconomic analytical techniques, suggest in Chapter 3 of this book that the profitability of conversion is about half that found in the civilian sector, and that the overall economic contribution of conversion is less than 2 percent of total industrial production value, even as the percentage of converted products in the MIC's output has continued to rise. Clearly there are no superprofits here; rather, these results suggest that the social imperative is paramount even if China's leaders have military modernization on their minds.

The sector's managerial problems are further reflected in press comments about the need to "break down closed-style management" and in calls to bring the defense industry "into the orbit of the national economy." As *China Aviation News* noted in a comment published in 1993 (which is typical of the dialogue):

> With new socialist market economic structures being established in China, all enterprises should go to the market and participate in market competition. Military enterprises have not adapted well, [and] must reform the existing military industrial mechanism, set up new management models . . . [and] change old concepts that military enterprise can only practice planned economy.[51]

Other authoritative critiques of Chinese defense conversion have been voiced. Electronics Minister Hu Qili has criticized the electronics industry for its backward technology, inadequate economies of scale, poor product quality, inadequate variety, unrealistic prices, and deficient customer service.[52] At a conference on arms control held in Beijing in 1994 that I attended, Chinese defense scientists and engineers voiced the following criticisms of the Chinese defense conversion effort:

- Lack of capital for conversion
- Low quality and low technology of defense conversion products
- Uncompetitive pricing of defense conversion products
- Poor "product structure" arising from lack of market research (that is, defense conversion products are market chasers, copy what is already on the market, and are uncompetitive)

- Inability to achieve economies of scale in producing for the market
- Lack of customer consciousness and lack of understanding of market operations
- Poor financial shape of the defense industry, which is plagued by the problem of triangular debt and nonrepayment of the loans made by the state to bail it out of financial problems
- Lack of cost consciousness
- Drag on the entire effort from subsidies required for converting the Third Line
- Management and organizational problems: Managers still have a planned economy mentality, depend too much on the state, and are unable to adapt to market demands
- An inappropriate approach to defense conversion: Leaders look for a technological quick fix, whereas the problem is really one of corporate culture[53]

All of these criticisms come from Chinese, not foreign, sources. Yet many are typical of remarks made about conversion and MICs around the world. They reflect problems not only of defense conversion but also of conversion of a command economy to one responsive to the marketplace.

The View from the Shop Floor

As long as we stay at the strategic or policy levels, we cannot really grasp the issues facing the defense conversion effort in China—we need to look at the process from the managerial point of view. Here, the inherent contradictions in the Chinese approach are laid bare. According to *China Defense Conversion News,* many defense plants are unwilling to continue to supply the military. One defense industry manager complained that he was caught in a bind, trapped between the new environment brought about by the economic reform program and the requirement that he keep military production going: His factory had to "expend 50 percent of our efforts to produce barely 5 percent of our output [military products]."[54] Yet at the same time, the OECD Development Center's conversion report indicated that most defense plants did not want to cut the cord with the defense industry bureaucracy.[55]

Other problems can be glimpsed from a paper about joint ventures for conversion presented by a NORINCO official at a conference on conversion held in Hong Kong in 1993.[56] The paper discussed seventy joint ventures with a total investment of U.S.$196 million; outside investors supplied 45 percent, or U.S.$88.6 million, of the base. JVs, the official noted, helped to change "product structure," to change the "closed style" of management, and to solve the capital shortage problem.

Of the seventy JVs, however, only fifty were in operation; twenty-nine made a profit of U.S.$9.24 million and twenty-one had losses of U.S.$7.64 million, thus leaving a positive balance of U.S.$1.6 million. But there were problems. First, there was the issue of "one factory, two systems"—because the JV made money and the remainder of the plant lost it, the JV got preferential treatment from management. The relationship was not mother-son (which was the intention) but rather one of competing brothers. Second, there was the issue of "two ends on the outside"—it appears that these JVs were primarily low value-added processing agreements, which meant essentially that imported inputs were expensive (and cost foreign exchange), but the outputs were low priced.

In fact, if we look at the numbers more carefully, we can see that over the eight years the average income from profitable factories was U.S.$318,500, whereas on the negative side the average annual losses were around U.S.$364,000. The average rate of return on foreign investment was 1.8 percent. This analysis is very crude—we would need much more detailed information to go farther, data that in China is apt to be secret—but it does coincide with what we know about most JVs in China: They are small. It also corroborates reports about loss makers in the state sector. In sum, these results would appear to be disappointing. Indeed, speaking privately, a high NORINCO official told a visitor in early 1995 that defense conversion in China was problematic—the few successes were vastly outnumbered by problems and potential bankruptcies. The greater part of the industry—"90 out of 100," he said—had problems meeting payrolls. The "successes" of conversion were exaggerations; the regime was "not telling the whole truth."[57]

Future Issues

The future of defense conversion in China depends on the future of the larger reform program in China. Along with efforts have come problems such as opportunism, profiteering, and corruption. The MIC, more specifically, faces personnel problems. Reports published in late 1992 said that one-third of all scientists in Shanghai were unemployed and another third were only partially employed; a Chinese Academy of Sciences survey indicated that 90 percent of China's current research professionals would be in retirement by the year 2000.[58] It is hard enough to keep people in the remote Third Line sites, let alone to attract young graduates there.[59] As one military scientist said, he would be willing to serve his country in the hinterland but could not accept the poor opportunities that location would afford his children.

It seems clear that military needs are hardly being ignored. The question of whether Chinese military expenditures are increasing, stagnant, or even declining has sparked a growth industry among analysts.[60] But regardless of the actual status of these expenditures, the Chinese press continues to print both calls for "self-reliance" and "strengthening leadership in military production" on the one hand and reports of research, development, and deployment of modern weapons systems and discussions of military strategy and army building "in the new era" on the other.

Regardless of what one may think of the current Beijing regime, the way it defines its national security goals, and its behavior in the South China Sea or toward Taiwan, all states have legitimate defense needs. It makes sense to conserve and rebuild those operations that have a chance of supplying China's new military needs. The much larger number of factories and enterprises that cannot contribute could be cast off to sink or swim in the sea of the market economy. This indeed would be conversion by total immersion.

At the same time, we should not forget that the Chinese defense conversion effort is embedded in the larger Chinese economy. The advances and problems outlined above take place across the system. The defense sector and its conversion are subject to the effects of the Chinese business cycle. According to the official State Statistical Bureau (SSB), there have been four such cycles since 1979, driven by inappropriate use of policy controls on prices and investment. The SSB, in a year-end review of the economy in 1994 published in *People's Daily,* said that "the problems of [the state sector] developing relatively slowly and lacking vitality are . . . quite conspicuous." By the end of the first eleven months of the year, 41.4 percent of the "30,000 enterprises included in the [state] budget" had lost money, and the amount lost (not specified) was almost 28 percent greater than it had been in 1993. The losers were in virtually all sectors of the state economy, but among the "greatest losses" was the military. Problems of management, unpaid long-term debt, poor product quality and mix, a growing amount of unsaleable goods, and the social burden were blamed for the poor performance. Losses in the defense sector were "mainly caused by historical reasons"—a vague term that must include the legacy of the Third Line.[61]

Thus, it is not surprising to find that Chinese observers rank moving from "a condition of high costs and low profits" as a key task. The opening chapter of *Chinese Defense Conversion,* attributed to COSTIND conversion chief Huai Guomo, calls upon the defense sector to shift to business management practices, form enterprise groups, make better use of the sector's technological capabilities, develop international links, and upgrade personnel.[62] CAPUMIT Vice Chairman Jin Zhude suggested that success in defense conversion includes[63]

- Dealing with problems of social security for the defense sector
- Transforming management and enterprise organization and governance to meet the demands of a market economy
- Moving from capital-intensive to technology- and information-

intensive industry that meets international standards of accounting and quality control

- Developing "mainstay" or "pillar" industries that can drive development, such as the automotive sector
- Coupling conversion with regional economic development plans
- Ratcheting up the scale of production and developing financial capabilities
- Developing international cooperation, particularly by introducing technology and capital from abroad
- Achieving success in the international marketplace

This laundry list of tasks applies equally to the state sector, of course; it is also clear that even if we grant the conversion process in China some success, the road ahead will be arduous. But authoritative statements also suggest that time is getting short. Huai Guomo calls conversion "a second beginning" (*dierci chuanye*) for the defense sector.[64] Jin Zhude notes that economic austerity moves may have weakened conversion: "Since 1993, as China strengthened macroeconomic controls . . . the development momentum of conversion has clearly weakened, and it is necessary to make new restructuring and plans." He also suggests that defense conversion is critical to continuing reform, especially given the new Ninth Five-Year Plan's emphasis on the development of China's interior. Indeed, in an echo of the conversion dialogue noted above, there is more than a hint of crisis in his urgings that the defense sector must respond to central directives—otherwise, it "again could become a major burden on or hindrance to [*baofu,* a very strong word with negative implications] China's economic development, as it was in the 1970s."[65]

In sum, the outcome of defense conversion in China is decidedly mixed. The difficulty of matching practical results with strategic intentions is certainly one of the lessons of the Chinese experience. Jörn Brömmelhörster, in his concluding chapter comparing Chinese conversion efforts with those in Russia, goes further in pointing out the systemic difficulty of transition and reform in socialist economies. Another lesson, he suggests, is that even modest success in defense conversion depends on the presence of a sound civilian economy.

The analysis we have presented here, like all analyses, is based on certain assumptions about the PRC and its leadership: the primacy of politics for Chinese decisionmakers; their desire to restore China's historical greatness and to make China a great modern power; and their perception that China exists in a unstable, hostile international environment in which China is a rising force that is being opposed by declining, but still powerful, nations. It should be clear that the decision to move toward defense conversion in China is hardly part of the detailed, well-thought-out plan by a farseeing leadership sometimes portrayed in the Chinese press and elsewhere.

It would be more accurate to describe it as an uneven and perhaps contradictory response to the ad-hoc program of economic reform launched in the late 1970s, itself an improvisation driven by the post–Cultural Revolution political requirement to relegitimize the Party. If, as the Chinese expression goes, reform is like crossing a stream while feeling the stones with one's feet, then in one's meandering process across the water it is inevitable that toes get stubbed and ankles twisted. Indeed, as the contributors to this book suggest, reform and conversion have unleashed conflicting forces: maintaining a labor force or building economically efficient enterprises, building national defense or contributing to the larger economy. At that, one should not expect China's road to wealth and power to be easy.

Notes

The author would like to acknowledge the considerable assistance and encouragement given him by Bates Gill of the Stockholm International Peace Research Institute (SIPRI); Lieutenant Colonel Dennis Blasko, U.S. Army; Peter Almquist, U.S. Arms Control and Disarmament Agency; Lynne Rienner and her staff; an anonymous reviewer; and, in particular, the coeditor, Jörn Brömmelhörster. Some of this material appears in different form in John Frankenstein and Bates Gill, "Current and Future Challenges Facing Chinese Defence Industries," *China Quarterly* (special issue on the PLA) 146 (June 1996), and is used with permission. Any errors, omissions, or misjudgments are, however, the author's, who would like to hear about them.

1. Actually, the current round of conversion is counted by Chinese authorities as the fifth time the military industries have changed their strategic direction. According to one Chinese source, the first shift—toward civilian efforts—came immediately after 1949, when the new regime consolidated the remnants of the old Guomindang defense industrial complex; the second, toward war production, occurred during the Korean War; the third—back toward an emphasis on civilian production—started in 1953 during the First Five-Year Plan, when backbone enterprises were being established with Soviet help; the fourth shift, again toward military production, came in the 1960s and 1970s when relations with India and the Soviet Union worsened and U.S. involvement in the Vietnam War deepened; and, finally, the fifth occurred when China and Vietnam fought a border war. See Wang Li et al., eds., *Dangdai Zhongguo-de Bingqi Gongye* in the series *Dangdai Zhongguo Congshu* [The Contemporary Chinese Ordnance Industry] (Beijing: Dangdai Zhongguo Chubanshe, 1993), pp. 384–386. In other words, in the Chinese view, "conversion" can swing both ways—from war production to civilian production and back again. This is consistent with the Chinese view that war and peace are—like *yin* and *yang*—two sides of the same coin, and thus, conceptually at least, the dual role of the defense industry in the larger economy and the relationships between the civilian and military sectors suggested by the "16-Character Slogan" are without contradictions. For other accounts of Chinese defense conversion, including a fascinating Marxist-Leninist apology for the introduction of market forces into the defense sector plus industrial and enterprise cases, see Cao Shixin, ed., *Zhongguo Jun Zhuan Min* [Chinese Defense Conversion] (Beijing: Zhongguo Jingji Chubanshe, 1994); see also Paul Humes Folta, *From Swords to Plowshares? Defense Industry Reform in the PRC* (Boulder, Colo.: Westview Press, 1992). For broader accounts of the

Chinese defense sector, see John Frankenstein, "The People's Republic of China: Arms Production, Industrial Strategy and Problems of History," in *Arms Industry Limited,* ed. Herbert Wulf (Oxford: Oxford University Press/SIPRI, 1993); John Frankenstein and Bates Gill, "Current and Future Challenges Facing Chinese Defense Industries," *China Quarterly* (June 1996); and Wendy Frieman, "China's Defence Industries," *Pacific Review* 6:1 (January 1993).

2. Some authors use the term "Third Front" to describe these industries.

3. See "Conversion of Military Technology to Civilian Use Discussed," *Foreign Broadcast Information Service, Daily Report China* (Washington, D.C., hereafter *FBIS-CHI*), 25 March 1994, p. 35.

4. China Electronic Industry Trading Delegation catalog, 1993 Hong Kong Conference on International Cooperation to Promote Conversion from Military to Civilian Industry, Hong Kong, 7–11 July 1993, p. 1.

5. See "'War Industry' Produces More Civilian Products," in *FBIS-CHI,* 9 June 1994, p. 36.

6. See "Arms Maker Produces More Civilian Goods," in *FBIS-CHI,* 19 January 1995, p. 36.

7. Paper distributed at OECD-sponsored International Conference on the Conversion of China's Military Industries, Beijing, 26–27 June 1995. A European diplomat noted privately—and with some amusement—that many of the high-ranking Chinese attending the conference whom he knew as flag-rank PLA officers were introduced to the conference's foreign participants as "Mister"—a real example of "defense conversion."

8. Project Manual for International Investment and Collaboration, Trade Delegation of the China Aviation Industry, 1993 Hong Kong Conference on International Cooperation to Promote Conversion from Military to Civilian Industry, Hong Kong, 7–11 July 1993, p. 2.

9. For such a view, see "Reforming the Pentagon: An Inside Job" (interview with J. M. Deutch), *Technology Review* 97:3 (April 1994), p. 31.

10. See J. T. Lundquist, "Shrinking Fast and Smart," *Harvard Business Review* (November-December 1992).

11. Folta, *From Swords to Plowshares?* p. 1.

12. For a sampling of literature that suggests the intractability of the issue, see Bruce D. Berkowitz, "Why Defense Reinvestment Won't Work," *Technology Review* 97:5 (July 1994), pp. 52–60; John Frankenstein, "Dilemmas of Public Policy and Strategic Management: The Conversion of Defense Industries to Civilian Production," *The International Executive* 37:2 (March-April 1995), pp. 105–125; and the critiques in Jacques Gansler, *The Defense Industry* (Cambridge, Mass.: MIT Press, 1980).

13. See "GDP up 10.2%, meeting target," *China Daily,* 6 January 1996, p. 1.

14. See John Wilson Lewis and Xue Litai, *China's Strategic Seapower: The Politics of Force Modernization in the Nuclear Age* (Stanford: Stanford University Press, 1994), p. 76, citing COSTIND publications and a commentary by Ivan Arkhipov.

15. See Lewis and Xue, *China's Strategic Seapower,* chapter 4, for an account of the politics, often driven by personal animosities fueled by the maneuverings of elite survival politics of Mao's court, that led to the "bewildering array of bureaucratic organs" of MIC organization. See also Frankenstein, "Arms Production."

16. See Bates Gill, "Defensive Industry," *Far Eastern Economic Review,* 30 November 1995, p. 62. The move was prompted by fears of possible U.S. attacks arising from Chinese assistance to North Vietnam. According to documents released by the Woodrow Wilson International Center for Scholar's Cold War History Project,

the Chinese feared that their industry was overconcentrated on the vulnerable eastern seaboard—a report by then PLA Deputy Chief of Staff Yang Chengwu noted that "industry is over-concentrated . . . 52 per cent of the national defense industry (including 72.7 per cent of the aircraft industry, 77.8 per cent of the warship industry, 59 per cent of the radio industry, and 44 per cent of the weapons industry) are concentrated in 14 major cities." Not only were the centrally controlled industries thus involved in this massive move to the "big" Third Line, but also the provinces were urged to build their own "small" Third Line for self-sufficiency in small arms and ammunition manufacture. See Qiang Zhai, "Beijing and the Vietnam Conflict, 1964–1965: New Chinese Evidence," Cold War International History Project Bulletin special issues on "The Cold War in Asia" (draft version), Washington, D.C., winter 1995–1996, pp. 233–250; Yang's document begins on p. 243.

17. The authoritative article on the Third Line is by Barry Naughton, "The Third Front: Defence Industrialization in the Chinese Interior," *China Quarterly* (September 1988), pp. 351–386. See also Lewis and Xue, *China's Strategic Seapower,* chapter 4.

18. See "Third Line Military Enterprises Expand Operations," in *FBIS-CHI,* 21 July 1992.

19. Chen Daisun, ed., *Zhongguo Jingji Baike Quanshu* [Encyclopedia of the Chinese Economy] (Beijing: Zhongguo Jingji Chubanshe, 1993), p. 1754. Folta also presents some statistics derived from 1980s data that are roughly consistent with these figures. Folta, *From Swords to Plowshares?* p. 274.

20. Even in interviews given on a nonattribution, background basis, Chinese defense industry officials are coy and unresponsive when asked for specifics. How China expects to attract foreign investment to this sector—a seeming priority—without divulging industrial basics remains unclear. Press reports, such as those cited throughout this chapter, only give a partial view.

21. *China Aviation News,* 2 June 1994, cited in *Defense Science, Technology and Industry Digest* (*DSTI*), Hong Kong, June 1994, p. 5.

22. "Shaanxi Develops Science and Technology Industry from Defense Enterprises," in *FBIS-CHI,* 23 September 1993, p. 39.

23. In constant 1990 U.S. dollars; source: SIPRI arms trade database, 1995.

24. See Ken Allen and Richard Latham, "Chinese Defense Reform: The Air Force as a Case Study," *Problems of Communism* (May-June 1991).

25. The degree of involvement of combat units in commercial operations is unclear. In 1994 there were news reports suggesting that combat troops had stood down from commerce, but given the requirement that the army continue to supply much of its own support, the pressure to make money remains. In March 1994, *Jiefangjun Bao* [Liberation Army Daily], the official army paper, reported that "combat units below army level are no longer engaged in business production, the financial reserves at their disposal have correspondingly decreased, and the contradiction between fund supply and demand has become even more apparent." See "Circular Urges Easing Troops Financial Burdens," in *FBIS-CHI,* 4 April 1994, p. 42. The PLA also has problems arising from the general social relaxation that has accompanied the reform process—the appeal of "green wine and red lights" has led to attempts to strengthen discipline by restricting the troops to bases.

26. *China Daily,* in *FBIS-CHI,* 14 April 1987, pp. 31–32.

27. China Xinxing Corp. (Group) brochure, distributed at the 1993 Hong Kong Conference on International Cooperation to Promote Conversion from Military to Civilian Industry, Hong Kong, 7–11 July 1993.

28. "Military Organizes Transnational Enterprise Group," in *FBIS-CHI,* 9 December 1992, p. 33.

29. Handbook of the Chinese Aviation Industry Trade Group, distributed at the 1993 Hong Kong Conference on International Cooperation to Promote Conversion from Military to Civilian Industry, Hong Kong, 7–11 July 1993.

30. An observer at a conference on Chinese defense conversion held at Stanford University in 1993 noted that in conversion efforts "the Chinese [are] most interested in those technologies not only where China [is] weakest, but also where the U.S. is strongest: e.g., systems integration, avionics electronics, software."

31. My thanks to Bates Gill of the Stockholm International Peace Research Institute for raising this point. My discussions with Chinese officials and managers on the matter of coming clean on statistics leads me to conclude that (1) the issue is sensitive; (2) some do not understand why outsiders—including potential investors—would require accurate information; and (3) some simply do not grasp the analytical issues at all. There is also the matter of the accuracy of any Chinese figures—the State Statistical Bureau often complains about politically inflated reporting. Paul Folta devotes the good part of a chapter discussing these issues; see Folta, *From Swords to Plowshares?* chapter 5, especially pp. 111–127.

32. Defense plant managers interviewed by Paul Folta said they expected profits from civilian production to aid defense production modernization. Conversion activities would also allow military production costs to be spread over a wider base. See Folta, *From Swords to Plowshares?* p. 168.

33. In *Jiefangjun Bao,* 25 and 28 September 1993, cited in *DSTI,* 9 September 1993.

34. "Military Organizes Transnational Enterprise Group," in *FBIS-CHI,* 9 December 1992, p. 33.

35. Unpublished papers by Zhang Hongbiao of the Aerospace Ministry and E' Wanyou, Vice Director, Hubei Economic Commission, circulated at the International Cooperation in Peaceful Use of Military Industrial Technology Conference, Beijing, 1991.

36. "Liu Huaqing Urges Development of Defense Technology," in *FBIS-CHI,* 30 January 1995, p. 30 ff.

37. "Civilian Firms Help in Military Hardware Production," in *FBIS-CHI,* 25 January 1995, p. 24.

38. Cited in Xiang Wang, "Development of Modern Technology and Defense Conversion: Interview with Huai Guomo, deputy director of the Commission of Science, Technology and Industry for National Defense," *Conmilit,* no. 296 (May 1993), p. 4. This perception must have been reinforced when Huai was a visiting fellow at the Stanford University Center for International Security and Arms Control, located in the heart of Silicon Valley, in 1993. In any case, some of COSTIND's joint ventures with U.S. firms, such as Hua Mei Telecommunications, may have been set up with "spin-on"—as well as minimization of potential problems with export controls—in mind. See Bruce Gilley, "Peace Dividend," *Far Eastern Economic Review* 11 January 1996, pp. 14–16.

39. Unpublished paper by Li Hechun, China Defense Science and Technology Information Center, circulated at the International Cooperation in Peaceful Use of Military Industrial Technology Conference, Beijing, 1991.

40. Jin Zhude and Chai Benliang, "Strategic Thinking of China's Conversion in the 1990s," in Proceedings of the International Conference on International Cooperation in Peaceful Use of Military Industrial Technology, ed. Chai Benliang and Zhang Junping (Beijing: International Academic Publishers, 1992.)

41. T. Poole, "China's Army Storms Hong Kong with Goods for Sale," *Independent,* July 1993.

42. Jin and Chai, "Strategic Thinking."

43. This view has been expressed to the author personally by both Chinese and foreign observers. CCP Chairman Jiang Zemin put the contradiction between economic development and stability at the top of his "12 Major Relationships" issued at the end of the Fifth Plenum in September 1995. The matter exercises Chinese across the political spectrum, from democrats to "Third-Eye" conservatives. See the useful summary in Steven Mufson, "Chinese Reforms: 'Neoconservatives' Cast a Cold Eye—Challenges to Deng Flow into the Mainstream," *International Herald Tribune,* 14 November 1995, p. 2; and Chen Yizi, "A Realistic Alternative for China's Development and Reform Strategy: Formalized Decentralization," *Journal of Contemporary China,* no. 10 (fall 1995), pp. 81–92.

44. "State Culprit for Surge in Company Losses" (Reuters), *South China Morning Post (Business),* 15 November 1995, p. 4.

45. See Tai Ming Cheung, "On Civvy Street: China's Lumbering Arms Makers Face Market Rigours," *Far Eastern Economic Review,* 6 February 1992, pp. 40–43.

46. See "Weaponry Industry Improving Profit Margin," in *FBIS-CHI,* 23 January 1995.

47. Jean-Claude Berthélemy and Saadet Deger, *Conversion of Military Industries to Civilian Production in China: Prospects, Problems and Policies* (OECD Development Center Report, draft, May 1995), p. 26.

48. According to *China Daily Business Weekly* in May 1995, the state had invested 1,000 billion yuan in "technical upgrading" in the period 1978–1994; in 1994 alone the amount came to 40 billion yuan. Cited in *FBIS-CHI,* 8 May 1995, pp. 50–51.

49. Reports in March 1995 indicate the perilous finances of the state-owned enterprises: By the end of 1994, industrial debt (that is, money owed by factories to each other within the sector) had reached 631 billion yuan, and unsold inventory had reached a value of 401 billion yuan. *South China Morning Post* (7 and 10 March 1995) and *China Daily* (5, 6, and 12 March 1995), cited in *DSTI,* March 1995, pp. 2–3. In terms of percentages, these negatives represent:

	yuan	% of GDP	% of Industry Value Added
1994 Industrial Debt	631 billion	14%	34%
Unsold Inventory	401 billion	9%	22%

(Source of GDP and value-added figures see *Jingji Ribao* [Economics Daily], 1 March 1995.)

50. *Renmin Ribao* [People's Daily], in *FBIS-CHI,* 7 November 1991, "Conversion," p. 32; Berthélemy and Deger, p. i.

51. Cited in *DSTI,* June 1993.

52. *China Electronics Daily,* 9 June 1993, in *DSTI,* June 1993, pp. 8–9.

53. Interestingly enough, the same general comments about the state of Chinese defense conversion are also made in an essay by Jin Zhude, the vice chairman of CAPUMIT, titled "The Development and Policy of Chinese Defence Conversion," Beijing 1995, which was circulated at the OECD-sponsored International Conference on the Conversion of China's Military Industries, held in Beijing, 26–27 June 1995, and at meetings on defense conversion in Chongqing and Wuhan, 30 October–9 November 1995.

54. *Zhongguo Junzhuanmin Bao* [China Defense Conversion News], 2 April 1993. My thanks to Evan Feigenbaum of Stanford University for drawing this news clip to my attention.

55. Berthélemy and Deger, *Conversion,* p. 59. This result was obtained from factories in Yunnan, where most of the MIC is under local control.

56. Gu Wei, "Sino-Foreign Joint Venture and Cooperation—An Important Approach to Transference from Military to Civil Production," paper presented at the 1993 Hong Kong Conference on International Cooperation to Promote Conversion from Military to Civilian Industry, Hong Kong, 7–11 July 1993.

57. Two papers discussing the Jianshe motorcycle joint venture between NORINCO and Yahama presented at the OECD-sponsored International Conference on the Conversion of China's Military Industries, held in Beijing, 26–27 June 1995, mentioned production accomplishments but not profits; a visitor to the Jianshe cooperation in October 1995 reports that profits, like management relationships in the joint venture, are not clear. The paper by the Japanese executive in charge of Jianshe notes many operational problems that would be familiar to experienced China traders: problems arising from lack of foreign exchange, quality issues, disputes over technology valuation, language barriers, and "differences in the ways of thinking and habits." Or, as another experienced observer has put it privately, the biggest problems in dealing with Chinese joint ventures have to do with the conceptualization of management and engineering: "hammer and chisel engineering in the age of CAD/CAM."

58. *Science and Technology Daily,* 13 November 1992, cited in *DSTI,* November 1992, p. 3.

59. *Junzhuanmin Bao* [Defense Conversion News], 22 June 1993, cited in *DSTI,* June 1993, p. 18.

60. See Ka Po Ng, "China's Defense Budgeting: Structure and Dynamics," in *China Review 1995,* eds. Lo Chi Kin et al. (Hong Kong: Chinese University Press, 1995) for a useful discussion of the Chinese military budget and the analytical issues it poses. There are substantial problems of analysis in the matter of Chinese military expenditure: what is covered by the amounts openly described as the military budget, "off-budget," and hidden expenditures, etc. In nominal terms, the budget has seen major increases since 1989; in 1995, the published figure reached 63 billion yuan, up more than 21 percent over 1994.

61. "Statistics Bureau on 1994 Economic Situation," in *FBIS-CHI,* 3 February 1995, p. 32; see also "Economists Review 1994, Project 1995 Economy," in *FBIS-CHI,* 3 May 1995, pp. 35–41.

62. Cao, *Zhongguo Jun Zhuan Min,* pp. 7–9; *FBIS-CHI,* 27 July 1995, supplement, pp. 7–8.

63. The section summarizes arguments from Jin's 1995 OECD/Chongqing conference paper. Development of the automotive sector will be one of the key themes of the Ninth Five-Year Plan; according to a 20 August 1995 report in *China Daily,* the Ministry of Machine Building plans to pump 100 billion yuan into China's eight leading automotive enterprises over the next five years. As all of these enterprises can no longer be counted as part of the MIC, this projected central investment should not be seen as support for conversion but rather as part of the broader marketization push. For a detailed account of how one of the leading automotive enterprises escaped from the MIC, see William A. Byrd, "The Second Motor Vehicle Manufacturing Plant," in *Chinese Industrial Firms Under Reform,* ed. W. A. Byrd (New York: Oxford University Press/World Bank, 1992), pp. 371–425.

64. Cao, *Zhongguo Jun Zhuan Min,* p. 10.

65. Jin, in OECD/Chongquing conference paper entitled "The Development and Policy of Chinese Defence Conversion," Beijing, 1995. In the Chinese text this is *youyou keneng chengwei yinxiang guomin jingji fazhan de baofu, zhongjian 70 niandai de lishi.*

2

Conversion and National Security

Kay Möller

China's conversion of defense industries since the early 1980s has its roots in several developments, including economic reform (reorienting idle capacities), change in the international and regional environments (the "peace dividend"), attempts at military self-sufficiency, and both the promotion of import substitution and, conversely, the facilitation of dual-use imports through joint ventures and other activities.[1] Whereas some of these motives appear to reinforce each other, others seem to be mutually exclusive in that they emphasize aspects of either armament or disarmament. In this chapter I discuss whether conversion in China is part of a deliberate strategy of either disarmament or, on the contrary, military modernization. Toward this end, I examine the macro level of national security—threat perceptions, resulting doctrine and strategy since 1979, and relevant points of view within the political and military establishment.

Security Environment

Viewed objectively, China's external security environment has varied enormously since the late 1970s when conversion became a part of Chinese national policy. Although it may seem paradoxical that China would begin conversion under unfavorable security conditions, it seems clear that the leadership bet that domestic economic problems arising from a moribund command economy—stagflation, low quality of life, declining technologies—were a greater threat than potential foreign perils. It is useful, therefore, to review the international situation China faced.

During the 1960s and 1970s, China's most critical security threat came from the Soviet Union, which caused about two-thirds of the People's Republic of China's (PRC) military forces to be routed to the north and northwest. Border skirmishes across the two countries' Manchurian border in 1969 forced China to begin a slow rapprochement with the United States.

The threat abated, however, when Mikhail Gorbachev unilaterally decreed a reduction in Soviet far eastern forces and a withdrawal from Afghanistan in 1988, and terminated Soviet support for Vietnam's intervention in Cambodia. Normalization officially occurred with Gorbachev's visit to Beijing in June 1989. Nevertheless, it should be remembered that China had responded positively to certain Soviet overtures as early as 1981 and, disappointed with the development of its "strategic partnership" with the United States, had proclaimed a policy of equidistance vis-à-vis both superpowers the following year. Subsequent bilateral negotiations on borders and other issues suffered setbacks before Gorbachev came to power, but Beijing's reaction to the demise of the USSR in 1991 clearly indicated a preference for a partnership between Leninist neighbors rather than the uncertainties presented by a postcommunist empire, which might unleash democratic, Islamic, and other forces in the process of fragmentation.

The 1980s: The Evolution of International Relations

When China's conversion effort began, Vietnam had moved into Cambodia, Moscow had invaded Afghanistan, and the Soviet conventional and strategic threat along China's borders persisted (the threat increased with the Soviet decision in the mid-1980s to deploy SS-20 missiles in Asia). At the same time, the Carter administration had stopped short of joining Beijing in an "international anti-hegemonistic" (read anti-Soviet) front, the U.S. Congress had reaffirmed Washington's security obligations toward Taiwan, and presidential candidate Ronald Reagan was campaigning on pro-Taiwanese topics. Irritation over Reagan's Taiwan policy—particularly U.S. arms sales to the island republic—continued well into 1983 and was a primary motive for China's declaration of an "independent foreign policy."

When conversion gathered momentum in the second half of the 1980s, the overall strategic situation had begun to improve for China: In August 1982, the United States committed itself to gradually reducing arms sales to Taiwan. In September 1983, Sino-U.S. strategic cooperation resumed, with Washington heeding Beijing's concerns in the framework of Intermediate-range Nuclear Force (INF) negotiations and defining the extent to which military technology would be sold to the PRC. In 1984, Ronald Reagan became the first U.S. president in ten years to visit China. In July 1986, Gorbachev, speaking in Vladivostok, signaled movement on such issues as the Sino-Soviet border and Afghanistan.[2] Moreover, the 1987 INF Treaty provided for nuclear disarmament in both Europe and Asia.

Gorbachev began to seriously implement his policies by late 1988, when he announced substantial unilateral troop reductions in Siberia and central Asia, agreed to a final withdrawal from Afghanistan, and pressured Vietnam into accepting a definite timetable for completing its pullout from Cambodia. In May 1991, a treaty was signed on the eastern sector of the Sino-Soviet border; Central Military Commission (CMC) Secretary-General

Jiang Zemin, on a visit to Moscow, came close to offering a Sino-Soviet relationship reminiscent of the 1950s.[3] Arguably, China's security environment had never been more advantageous than it was in the spring of 1989.

This situation took a turn for the worse—albeit more in subjective perception than in strategic reality—with China's post–Tiananmen Square isolation as well as with the collapse of international communism and of the Soviet Union itself. Deng Xiaoping blamed Tiananmen on "Western subversion" and on efforts to undermine Beijing's socialism with a strategy of peaceful evolution.[4] Abstaining in the UN Security Council's November 1990 vote on the use of military force against Iraq was one means of reaffirming China's status as an important global player, but the outcome of Operation Desert Storm contributed to PRC paranoia: The Beijing leadership had not expected a quick allied victory in the Persian Gulf and increasingly perceived the campaign as a first step toward U.S. hegemony in a unipolar world.[5] Concern over the Gulf War and its outcome would have been less marked without the simultaneous breakup of the Soviet bloc and the collapse of China's "strategic partnership" with Washington. When this process culminated one year later in the disintegration of the USSR itself, China envisioned the United States' potential use of such developments and encouragement of destabilizing forces along its periphery.[6]

To be sure, Russian president Boris Yeltsin reaffirmed all commitments made by Gorbachev and continued with border negotiations, leading in November 1992 to agreement in principle on further disarmament and confidence-building measures. Furthermore, Yeltsin visited Beijing the following month and signed a joint declaration that approached a mutual nonaggression pact.[7] Defense cooperation—including massive transfers of Russian technology—had been initiated under Gorbachev and may have played a role in negotiations by the ministers of defense in November 1993.[8] China has also considered encouraging Russian nationalism in an anti-American sense.[9] However, the border settlement and troop reductions remain incomplete, while Beijing continues to view central Asian nationalism and Islam with unease, pursue underground nuclear tests close to the Kazakhstan border in spite of a moratorium until recently observed by all other nuclear powers, and presumably tolerate—if not encourage—large-scale Chinese migration to Siberia. China has also failed to respond to repeated Russian proposals for northeast Asian security cooperation; Japan has increasingly been perceived as a potential military rival.[10] All this has occurred as Russia's Pacific Fleet is stranded in Vladivostok and the United States is reducing its western Pacific military presence.

Chinese Maneuvers

What has been China's response? In this paradoxical environment, the Beijing leadership has embarked upon paradoxical policies—conversion, military modernization, force reduction, and assertions of military dominance. Beijing has identified a historical opportunity to reestablish the leading regional role supposedly denied China by colonialists, "imperialists," and "social imperialists" since the Opium Wars. One aspect of this—and one particularly disturbing from a regional point of view—is Beijing's military modernization program, driven by a new naval strategy. According to official statistics, China's defense spending has tripled since 1988. Even allowing for 20–25 percent annual inflation, the actual budget may be as much as five times the official numbers—bringing it close to Japanese spending levels.[11] The PRC has begun to build a blue-water navy (with rumors of an aircraft carrier) and has modernized its nuclear forces. Included among the arms deals signed with Russia are fifty long-distance fighter bombers of the Su-27 Fencer type, which permit power projection in maritime areas far away from the Chinese mainland. Licensed manufacture of the aircraft in the PRC has been agreed upon in principle.[12] In the meantime, China has advanced in the South China Sea, has cultivated special relations with Myanmar, has tried to influence the process of Korean reunification, and has repeatedly threatened Taiwan with unification by force.

Whereas parts of the Chinese military establishment have pushed for a more aggressive stance in the South China Sea vis-à-vis Taiwan and the United States, the civilian leadership—caught in a power struggle for the Deng Xiaoping succession and considering the possible repercussions on economic development—opted for engaging regional neighbors and the United States in bilateral and multilateral security dialogues while at the same time strengthening the PRC's economic and military clout.[13] This approach appeared to pay off when President Bill Clinton detached trade from human rights considerations in May 1994. Military officials, although occasionally signaling flexibility in relations with stronger partners such as Washington (in nuclear and missile proliferation) and Moscow (with regard to the common border), have not indicated a readiness on behalf of the PRC to commit itself to more transparency—let alone disarmament—on a regional level. China's first-ever white paper on arms control and disarmament, issued in late 1995, did not go beyond an exercise in public relations.

In sum, the evolution of China's security perceptions between 1979 and 1995 provides a questionable rationale for conversion, which would seem to require the recognition of an objectively more benign international or regional environment. Beijing's evaluation of threats was rather unfavorable when conversion was first discussed between 1979 and 1983, justifiably positive when conversion yielded its initial results between 1983 and 1989, and pessimistic from 1989 to 1993. It has been caught between risk and opportunity ever since, with a regional great-power role in the making and third-party responses less predictable. To further examine the extent of the arms reduction element in China's conversion process, the internal strategic reply to perceived external developments must be discussed.

The Strategic Response

The Deng Xiaoping era began not only with calls for conversion, but also with a military campaign beyond China's borders—the "punitive expedition" against Vietnam in February and March 1979. In the course of that operation, the People's Liberation Army (PLA) occupied or besieged five provincial capitals near the border, destroyed Vietnamese urban infrastructure, and registered losses of several tens of thousands of soldiers.[14] If most Western analysts deemed the campaign a failure, it was largely due to the PLA's heavy reliance on tanks and superior numbers, whereas Vietnam's provincial and border troops could wage a guerrilla war on the rugged terrain. Neither side resorted to air support. Apart from various domestic, regional, and international considerations, Deng Xiaoping launched the punitive expedition as a test for a Chinese army that had last seen action in the late 1960s in North Vietnam (when supporting Hanoi) and in Manchuria (when engaging the Soviets along the Amur and Ussuri Rivers). According to Deng himself on the eve of the offensive:

> If a little disgrace is the outcome, we can judge our own power more clearly, we can observe whether we stand a test on a real battlefield with real rifles and guns, and which link in the chain of army work requires improvement. This has advantages for the modernization of national defense. Only when knowing one's own weak points and how to strengthen them, can we talk about building modern troops to protect the fatherland in the process of modernization.[15]

Consequences were outlined two years later by the commander of the Chinese Academy of Military Sciences, who recommended resorting to counteroffensive after positional warfare, combining different PLA branches to make maximum use of modern technologies, overhauling logistics, and phasing out "civilian" activities and political and ideological work.[16]

Meeting enemies at the gate—as opposed to drawing them into the interior—had actually been practiced since the Korean War, but the emphasis on technology, logistics, and communications was certainly new. PLA maneuvers throughout the 1980s centered on motorized infantry, air strikes, artillery, surface-to-air missiles, anti-tank missiles, pioneer units, and air transport. In 1985, Deng Xiaoping abandoned Mao Zedong's doctrine of the inevitability of global nuclear war, and the CMC followed up with a new emphasis on peacetime construction and preparation for local conflicts.[17] The introduction of tactical nuclear weapons signaled a new understanding of nuclear arms in a framework of conventional defense. Military planners urged the deployment of rapid reaction forces to "win small wars and keep medium wars under control."[18] Between 1985 and 1987, cuts in land forces led to an overall reduction from 4 to 3 million active soldiers. Also in 1985, the PLA navy drafted a plan to develop its coastal defense capacities into a blue-water force, including large warships, nuclear-armed submarines, air support, and aircraft carriers.[19] Modernization of both navy and air force has taken a decisive turn with high technology imports from Russia since the early 1990s.

The impact of the 1991 war in the Persian Gulf also worked as a catalyst for reform. China had initially expected the United States to become bogged down in prolonged ground battles and suffer a new Vietnam syndrome. Successful air operations and surgical strikes, however, vindicated those PLA commanders who had consistently underlined the importance of local warfare and modern weapons. The political leadership was now more inclined to listen to their advice.[20] In March 1991, PLA Chief of Staff Chi Haotian announced the deployment of rapid reaction forces as key elements of all combined group armies.[21] Special battalions were formed for amphibious, ground, and air operations.[22] In the field of nuclear defense, long-range delivery systems have been tested.[23] China's atomic tests also point to efforts at miniaturization.[24] Although strategic and doctrinal change since 1979 is proof of a shift toward a "leaner and meaner" PLA (and supporting defense industries)—with new emphasis on a high-tech navy and air force capable of long-distance power projection—it also signals new military objectives and a more active, if not aggressive, role rather than disarmament in recognition of a diminished threat.

Conversion, then, can be seen as part of a larger strategy of defense modernization. As Wendy Frieman has suggested, most of the 1980s conversion efforts affected facilities "that had actually been out of the weapons business for a long time."[25] At the same time, the navy and air force reaped the most profits from the transformation of military ventures into civilian business.[26]

The National Agenda and Power Struggles

Conversion may also be seen as a consequence of the debates and policies concerning military development in the post-Mao era. Conversion was presented as a way not only to advance the economy but also to modernize weaponry. I view this last pledge as an attempt by the regime to placate the PLA, which otherwise may have seen the commercialization of the civilian-run State Council defense industrial base as a threat. In this section I emphasize the moves made to accommodate the PLA and the army's increasing political role.

It is useful to remember that military modernization not only ranked last among Zhou Enlai's and Deng Xiaoping's "Four Modernizations" but also was accorded lowest priority in comparison with industry, agriculture, science, and technology during most of the 1980s. The share of the official defense budget in overall expenditure shrank from 17.5 percent in 1979 to 8.3 percent in 1987. Deng promoted professional officers at the expense of political commissars but denied their requests for high-tech arms imports until at least 1985. He insisted, instead, on the necessity of first accumulating capital as fast as possible—that is, through light industrial development. When possible, the PLA was encouraged to raise funds on its own through arms exports and civilian commercial activities.[27] With regard to State Council enterprises, it was decided that industries unable to produce critical hardware or to engage in arms exports would be converted to civilian production to provide both consumer goods and foreign exchange, which, in turn, would be used for high-tech imports.[28] This initially met with resistance in military and industrial circles[29] but nevertheless brought about instant results: Some authors allege that in the early 1990s, 65 percent of profits made from such activities were reported to have directly benefited the defense budget.[30]

China's own arms production had shrunk significantly during the 1970s, allowing excessive capacities to easily lend themselves to conversion. Just as the State Council ministries began to move into commercial operations, the PLA in 1984 initiated the conversion of airports, shipyards, ports, and railway tracks under its control to civilian use.[31] Usually this did not signify dissociation from the armed forces but rather the PLA's extension of its activities to industrial and other production so as to generate funds of its own. By 1991, about 2,500 products had been developed for civil consumption.[32] Most of the 800,000 workers then employed by about 10,000 military-related businesses were relatives of active soldiers.[33] This phenomenon was unavoidable insofar as the families of soldiers would have registered relative income losses had they not somehow participated in the overall opening up of the economy.

The problem of the PLA's extension of activities was partly due to the fact that Deng Xiaoping, while urging the military to "return to the barracks," actually needed to retain at least parts of it as a power base. His comeback in 1977 had been prepared by leading officers such as the commander of the Guangzhou military region, Xu Shiyou, and Deputy Minister of Defense Su Yu. Deng himself was chief of general staff and chairman of the State Military Commission until 1989. Troop reductions decreed in 1985, although serving purposes of rationalization, also facilitated the early retirement of officers opposed to the new policy.[34] Thus in 1980, eight out of thirteen commanders of military regions were replaced. Five years later, the number of military regions was reduced from eleven to seven, and the rotation of regional commanders was accelerated. In the autumn of 1985, seven high-ranking officers lost their seats on the politburo, to be replaced by civilians.[35] Yang Shangkun, the new secretary-general of the Central

Committee's Military Commission, was a Deng ally. Other loyalists were appointed minister of defense (Zhang Aiping), head of the PLA's Central Political Department (Yu Qiuli), and heads of the general staff and central political departments (Yang Dezhi and Yu Qiuli, respectively). Lower levels, including the military-industrial complex, were similarly affected.

Although military professionalism minimized the risks inherent in the emergence of such new power bases, the entire Dengist system became dependent on the functioning of a civilian leadership. This leadership obviously became paralyzed during the events leading to the 1989 Tiananmen Square massacre. When the PLA was summoned to resolve the crisis, a considerable number of field commanders and leading officers expressed their dissatisfaction—some even blatantly disobeyed orders.[36] The intervention resulted not only in a reemergence of the armed forces in national ideology and society as a whole but also in material rewards in the guise of promotions and a rise in defense spending. The 1990 defense budget registered an increase of 15.2 percent over 1989, marking the first real growth in ten years.[37] Between 1988 and 1995 defense expenditure tripled, even according to official statistics.

These increases were initially used to improve soldiers' living standards and to narrow the gap between their incomes and those of industrial workers.[38] The PLA gained more leeway for its commercial ventures to satisfy common soldiers and placate commanders. Earlier attempts at restricting such activities to agriculture, manufacture, and mining were abandoned.[39] It is unclear as yet whether this will lead to the creation of new military-civilian alliances in the provinces and thus possibly to warlordism;[40] the consequences for military ethics and national security are obvious, however, and of increasing concern to the central leadership.[41]

Conclusion

Deng Xiaoping has certainly not behaved as a modern politician who promotes civilian economic development with the aim of developing Chinese civil society or the PRC as a civil power. His point of departure has been the same as Mao's (or Hua Guofeng's, for that matter)—the goal of transforming China into a major power, thus claiming the rightful place supposedly denied to it since the mid-nineteenth century. He has been more successful than Mao in promoting this aim, because he understands the importance of internal and external stability in buying sufficient time for development. He has been less successful than Mao in that he was forced to partially surrender central control to achieve economic vitality. Internal and external complications became visible in 1989, but they further convinced Deng that one-party dictatorship could only be maintained by offering large segments of Chinese society the greatest possible material benefits.

Uncertain Consequences

Although conversion in the early 1980s may have been inspired by the overall motive of "civilian development first," it also contributed to maintaining the military-industrial complex and increasingly played a part in the general power struggle. It became even more prominent in this respect after 1989, due to domestic and external developments that Deng had unwittingly furthered or helped to further. Allowing the PLA to expand its industrial and support activities into the civilian market was motivated by a desire to reconcile China's armed forces with changes emerging in domestic and international society. At the same time, it enabled the PLA to once again develop into a potential political player. One may ask, therefore, whether conversion is the right term for this complex phenomenon at all. If anything, it has not worked in the sense of demilitarization and replacement of military options with peaceful means, but rather to the contrary.

Thus, what has been termed China's conversion of defense industries in the Deng Xiaoping era is anything but a deliberate strategy of disarmament. The evolution of China's security situation between 1979 and 1996 provides little evidence of replacement of the former Soviet threat by a secure strategic environment, and doctrinal and logistical shifts point to greater assertiveness in traditional zones of influence to the east and south. During the 1980s, conversion may actually have saved certain defense industries from becoming obsolete and the overall force from becoming a social problem. Imports of high-tech weapons from Russia, Israel, and other suppliers have since more than compensated for the phasing-out of outdated equipment and infrastructure. Where conversion extends into the domain of PLA commercial activities, it illustrates a desire to generate additional funds as well as concessions to military interests within the framework of the overall struggle for power.

Since Tiananmen Square, side effects have become more disturbing. The PLA may be more professional, better equipped, and depoliticized in an ideological sense. At the same time, however, it has clearly emerged as the single most important actor in both domestic and foreign politics. Conversion in this context is not the engine but rather one symptom of a process in which old elites are replaced with new ones. Viewed within the context of China as a "dissatisfied power," this combination may turn out to be an explosive one.

Notes

1. For a discussion of China's motives, see Mel Gurtov, "Swords into Market Shares: China's Conversion of Military Industry to Civilian Production," *China Quarterly* 134 (June 1993), pp. 213–241.

2. Speech by Mikhail Gorbachev at the Ceremonial Meeting Devoted to the

Presentation of the Order of Lenin to the City of Vladivostok, 28 July 1986, as quoted in *Far Eastern Affairs* (Moscow) 1 (1987), pp. 3–21.

3. *China Daily,* 8 June 1991, p. 3.

4. *Beijing Rundschau* 26 (11 July 1989), p. 28.

5. Harlan W. Jencks, "Chinese Evaluation of 'Desert Storm': Implications for PRC Security," *Journal of East Asian Affairs* 6 (summer/fall 1992), pp. 447–476.

6. Chinese politicians and political analysts have since then accused the CIA of destabilizing activities in Xinjiang, Inner Mongolia, Tibet, and Hong Kong. *Defense News,* 14–20 March 1994, pp. 1 and 37.

7. "Joint Declaration on the Basis of Relations between the PRC and the Russian Republic," *Renmin Ribao* [People's Daily], 19 December 1992, p. 1.

8. Paul Godwin and John J. Schulz, "Arming the Dragon for the 21st Century: China's Defense Modernization Programme," *Arms Control Today,* December 1993, pp. 3–8. The ministers denied allegations to that effect.

9. *Cheng Ming* (Hong Kong), 1 November 1993, pp. 20–21; *Lien Ho Pao* (Hong Kong), 21 October 1994.

10. Author's information.

11. David Shambaugh, "Growing Strong: China's Challenge to Asian Security," *Survival* 36:2 (summer 1994), pp. 43–59.

12. *Jane's Defence Weekly,* 6 May, p. 3.

13. In August 1993, "new and veteran Chinese generals" were reported to have petitioned President Jiang Zemin to "adopt appropriate measures against hegemonistic behaviour of U.S. authorities, which severely violate international law and China's national dignity and sovereignty" (*Ching Pao,* 5 October 1993). PLA meddling in these and other aspects of foreign policy was reported by *Cheng Ming* on 1 September 1994.

14. The PLA Deputy Chief of Staff mentioned "less than ten thousand" Chinese dead, but Hanoi counted 62,500. *Agence France-Presse* (Hong Kong), 2 May 1979, in *Foreign Broadcast Information Service (FBIS),* PRC E1-2, 3 May 1979; *Vietnam News Agency,* 19 March 1979, in *FBIS,* AP K28-31, 19 March 1979.

15. *Qishi Niandai* (Hong Kong), April 1979, pp. 25–26.

16. *Hongqi,* 16 August 1981, pp. 7–8.

17. Godwin and Schulz, "Arming the Dragon," p. 5.

18. *Jane's Intelligence Review,* August 1993, pp. 373–377.

19. You Ji and You Xu, "In Search of Blue Water Power: The PLA Navy's Maritime Strategy in the 1990s," *Pacific Review* 4:2 (1990), pp. 137–149.

20. Jencks, "Chinese Evaluation of 'Desert Storm,'" pp. 460–470.

21. *Jiefangjun Bao* [Liberation Army Daily], 28 April 1991.

22. *Jane's Intelligence Review,* August 1993, pp. 373–377.

23. *International Herald Tribune,* 13 October 1994, p. 2.

24. *International Herald Tribune,* 8–9 October 1994, p. 2.

25. Wendy Frieman, "China's Defence Industries," *Pacific Review* 6:1 (January 1993), p. 58.

26. In the late 1980s, the PLA navy participated in 5,300 construction projects, of which only 700 were directly related to national defense. The naval air force has been running eleven "dual-use" airports along the coast. François Guilbert, "L'Armée populaire de libération face à la régionalisation économique," *Relations Internationales et Stratégiques* 15 (autumn 1994), pp. 91–103.

27. Stacey Solomone, "The PLA's Commercial Activities in the Economy: Effects and Consequences," *Issues and Studies* (Taipei) 31:3 (March 1995), pp. 20–43. Wendy Frieman dismisses the argument put forward by the central government during the first phase of conversion that a transfer of "superior skills" from the

defense sector to the civilian economy was primarily meant to enhance the latter's productivity; Frieman, "China's Defence Industries," p. 57.

28. *Jane's Intelligence Review,* April 1993, pp. 186–189.

29. Gurtov, "Swords into Market Shares," p. 226.

30. Guilbert, "L'Armée populaire," p. 96.

31. *Jane's Intelligence Review,* April 1993, pp. 186–189.

32. *Jiefangjun Bao,* 12 December 1991, p. 1.

33. *China Trade Report* (Hong Kong), August 1993, pp. 5–6.

34. Plans drawn up in 1992 for a further demobilization involving 800,000 soldiers, among them 100,000 officers, initially met with resistance from the military leadership. *Cheng Ming,* 1 March 1992, pp. 12–13, and *Hsien Pao* (Hong Kong), 8 September 1992.

35. Among them was Xu Shiyou, who may have turned against Deng in the wake of the 1982 party congress.

36. Harlan Jencks, "Party Authority and Military Power: Communist China's Continuing Crisis," communication for the nineteenth Sino-American Conference on Mainland China, Taipei, 12–14 June 1990. General Yang Baibing, who lost his office as CMC secretary-general and head of the PLA's General Political Department in 1992, together with his half-brother Yang Shangkun (then first CMC vice chairman), was accused, among other things, of having neglected party work and political indoctrination within the armed forces. Sebastian Heilmann, "Die Armee und die Perspektiven der kommunistischen Herrschaft," *China aktuell* (January 1995), p. 27.

37. Stockholm International Peace Research Institute (SIPRI), *SIPRI Yearbook 1992: World Armaments and Disarmament* (Oxford: Oxford University Press, 1992), pp. 245–246.

38. *Jane's Defence Weekly,* 28 January 1989, p. 137.

39. Guilbert, "L'Armée populaire," p. 99.

40. Sebastian Heilmann mentions indications to the contrary, that is, increasing tensions among local administrations and military units, fueled by economic competition; Heilmann, "Die Armee," p. 31. This contrasts with François Guilbert's view that financial contributions by provincial authorities to PLA regional commands between 1984 and 1991 and favors granted in return have "slowly drawn the military into local configurations of political power." The most obvious example is Guangzhou, with at least 10 percent of all military businesses. Guilbert, "L'Armée populaire," pp. 92, 96, and 100.

41. *Jiefangjun Bao,* 26 July 1993. Recommendations made by the Academy of Military Sciences in March 1988 under the heading "One Army, Two Systems" implied that troops stationed in "unstable" regions should concentrate on their military duties rather than venture into business activities. *Jane's Intelligence Review,* April 1993.

3

An Economic Analysis

Feng-Cheng Fu & Chi-Keung Li

In this chapter we offer an analysis of the economic benefits of defense conversion to China based on official People's Republic of China (PRC) figures. Although these numbers are often questionable, some conclusions may be drawn from them. We begin with an outline of the main economic objectives of defense conversion policy and defense enterprises' conversion of production from military to civilian goods. Then, using official data, we examine the growth and current production scale of converting enterprises. We also estimate the direct and indirect contribution of defense conversion to China's economy. Some representative cases in the automotive industry are used to provide a microeconomic perspective. We then compare the performance of converted military enterprises to that of the whole economy and point out the advantages and disadvantages of military enterprises in the development process. Finally, we suggest some possible future directions for defense conversion in China.

Defense Conversion Policies and Implementation

Over the past fifteen years, civilian production has become a major activity for defense enterprises. According to the State Council, more than 15,000 items are now produced for the civilian market by the defense sector.[1] Along with production, however, the development of converted defense enterprises has included other mandates:

- *Undertake research and development (R&D) for important projects and key equipment* in order to fill in gaps in the country's scientific and technological fields and to allow China to reach First World levels. Such projects include the "863" and "Torch" scientific programs, which strive to develop new materials and electronics, as well as the Sichang Base, from which civilian satellites are launched.

- *Support national key industries, including energy and communications.* A large number of civilian products, such as aircraft, ships, automobiles, railway rolling stock and engines, and drilling platforms for oil exploration have been manufactured to fulfill the needs of the national economy. In particular, automotive manufacturing is the "pillar" industry for the defense sector. Among the eight big automobile manufacturing bases in China, three—Beijing Jeep, Chongqing Changan, and Guizhou Aviation—belong to the defense industrial system. Other important transportation equipment such as aircraft, ships, automobiles, and railway rolling stock and engines are also produced by defense enterprises. In the energy sector, some of the famous coal mines in Shanxi are operated by the People's Liberation Army (PLA). Other essential equipment for digging or extracting fuel, such as petroleum well-drilling platforms and various specialized vehicles for mining areas, are also researched and produced by defense enterprises. Official statistics indicate that the shares of output produced by the defense sector in total production are the following: automobiles, 9 percent; motorcycles, 60 percent; container trucks, 26 percent; and comprehensive digging equipment for mining, 24 percent.[2]
- *Supply various light industrial products and household electrical appliances to the home market.* During the surge in demand for home electric appliances in the 1980s, many defense enterprises invested in production facilities and became the primary producers of such items as cameras, television sets, and refrigerators.

Strategies

To execute these missions, defense sector enterprises have attempted varied strategies and organizational forms to achieve the transformations necessary for conversion. The principal schemes are the following:

1. *Combined military and civilian production.* This category may be classified into three subtypes, according to asset allocation:
   - Flexible production plants that can produce goods used for both the military and the civilian sectors. These enterprises can produce both types of products without additional inputs; after completing military tasks, they utilize idle capacity for civilian production.
   - Enterprises that retain dedicated military production lines. Using a portion of the original resources devoted to military products as well as new inputs, enterprises build production lines dedicated to civilian production and thereby achieve the purpose of integrating military and civilian products.

- Enterprises that not only maintain all defense production lines and continue to produce military goods but also invest in new equipment for civilian production. This is equivalent to investing in a new civilian factory.

2. *Production transfer.* When defense enterprises encounter lower military production orders, they may innovate and produce new civilian goods using current resources and production factors. They then become enterprises that produce only civilian goods but retain their status as defense enterprises under the defense industrial system.
3. *Enterprise dissolution and asset reallocation.* During the readjustment process of defense conversion, some defense enterprises are inevitably closed. All production transfers to other civilian enterprises through merger or asset transfer. Some factories are physically transferred to other regions.

Combined production seems to be the most common strategy adopted in the development and implementation of defense conversion in China. Very few enterprises can implement production transfer because the authorities do not allow the transferring enterprises to rid themselves completely of the duties of military production and the transferring enterprises are not familiar with civilian markets. Under these circumstances, transferring enterprises do not instantly abandon the military market and entirely concentrate on the civilian market; usually they must first experience the stage of combined production, and complete their transformation to civilian production after they are accepted by the civilian market. Because of considerations of stability, even fewer enterprises are approved by the authorities for the third type of transformation. Most of those that are being dissolved are Third Front (*sanxian*) defense enterprises that are taking the opportunity to move to the more developed coastal provinces.

Stages

The development directions and strategies of defense conversion have gone through several stages. During the Sixth Five-Year Plan (1981–1985), China adopted a strategy of laissez-faire with respect to conversion. Defense enterprises could freely choose the items they would manufacture for civilian production in order to meet perceived market needs. Bicycles, sewing machines, and watches thus became favorite products for military enterprises. By 1985, military enterprises had a production capacity of 3 million bicycles, 1 million sewing machines, 1.5 million laundry machines, 150,000 television sets, and 300,000 motorcycles.[3]

Nevertheless, every region overproduced civilian necessities in a frenzy of market chasing, and soon there was an oversupply of goods. Faced with production surplus and idle plants, China inevitably enacted a set of limiting regulations. Conversion therefore changed with the Seventh Five-Year Plan (1985–1990) as China began to fund the development of certain items based on an industrial policy of planned development. In 1986, the State Council and the Central Military Commission (CMC) approved 126 items for technical transformation to integrate military and civilian production. Out of a 1.6 billion yuan investment fund, 820 million yuan consisted of special loans to the defense sector. In 1988, the government issued the second list of transformation items. For a 3.4 billion yuan investment, the government arranged a 1.72 billion yuan bank loan. Furthermore, 400 million yuan were allotted to technological aspects of conversion in the Seventh Five-Year Plan. In the area of capital investment, the government also designated some core items for defense conversion.[4]

Converted enterprises experienced remarkable growth in output in the 1980s. The share of the output of civilian products produced by military enterprises rose from 6.9 percent in 1975 to 23 percent in 1980. During the late 1980s, this ratio increased to about 70 percent. In 1994, civilian production constituted 80 percent of total production of defense industry (see Table 3.1). Although there was rapid quantitative growth in their manufacture, however, the quality of civilian products was still less than

Table 3.1 Civilian Production as Percentage of Total Production in the Defense Industry

Year	Percentage
1975	6.9
1978	14.6
1980	23.0
1985	45.0
1988	66.0
1990	70.0
1992	76.0
1994	80.0

Sources: Committee of Editors, *Zhongguo Gongye Nianjian* [China Industry Yearbook] (Beijing: Zhongguo Laodong Chubanshe, 1991); China Machinery and Electronics News Agency, Military Technology for Civilian Use News, Zhongguo Jidian Baoshe, Jungong Jishu Minyong Baoshe, eds., *Zhongguo Guofang Gongye Jun Zhuan Min Jiehe Shizhounian Zhuanji* [Ten Years' Integration of Military and Civilian Production of China National Defense Industries] (Beijing: Zhongguo Jidian Baoshe, Jungong Jishu Minyong Baoshe, 1990); Xinhuashe [New China News Agency], 26 July 1995; Zhou Jianshe, *Guofang Ziyuan de Nixiang Kaifa—Zhongguo Jun Zhuan Min Wenti Yanjiu* [The Peaceful Development of Resources of National Defense—A Study of the Conversion of Military Resources to Civilian Fields in China] (Changsha, Hunan: Changsha Chubanshe, 1992), p. 204; Zhongguo Guowuyuan Xinwen Bangongshi [News Office, State Council of the PRC], "Zhongguo De Junbei Kongzhi Yu Chaijun" [Military Equipment Control and Military Reduction in China], *Da Gong Bao,* Hong Kong, 17 November 1995, p. A7.

Note: According to China's official definition, military industry includes armaments, aerospace, aeronautics, shipbuilding, and nuclear categories.

satisfactory. Some defense enterprises had turned into key producers of machinery and electronic products with relatively good economic results. Nonetheless, after embarking upon the course of economic construction, most defense enterprises encountered problems in entering the civilian market—poor information, inexperience, and keen competition. According to incomplete statistics, more than 30 percent of converted military enterprises could not achieve the requirement of responsibility for profit and loss and self-development.[5]

From Quantity to Quality

Under the Eighth Five-Year Plan in the 1990s, the policy of conversion focused on quality improvement, while quantitative growth was given second priority. Along with deeper economic reforms, institutional reforms of military enterprises became key facets of the plan:

- *Change in operational mechanisms and management.* Defense enterprises were authorized to make their own decisions on personnel, wage, and dividend distribution. Thus, they could respond to market changes, at least in theory.
- *Growth of corporations and enterprise groups.* Military enterprises could now integrate upstream and downstream enterprises to enjoy the benefit of economies of scale.
- *Development of a shared system of ownership,* thus promoting economic benefits of enterprises.
- *Adjustment of production and industrial structure.* By enhancing coordination between the State Planning Commission (SPC) and the Commission on Science, Technology, and Industry for National Defense (COSTIND), the tasks of military production were reduced to free more production capacity for civilian production.

The Impact of Conversion Policy on the Macroeconomy

Before discussing the impact of defense conversion on China's economy, we must examine the development of the defense industry with respect to national defense expenditures, the production value of converted enterprises, and the efficiency of these enterprises. What is the relationship between China's military burden and the development of defense industry conversion? We again draw from official statistics but subject them to our own analysis.

Defense expenditures. Figure 3.1 reveals that China's defense expenditures increased slightly during the 1980s. Compared to the overall growth of

Figure 3.1 Official Expenditure on National Defense

Source: State Statistical Bureau (Guojia Tongjiju,), eds., *Zhongguo Tongji Nianjian* [Statistical Yearbook of China] (Beijing: Zhongguo Tongji Chubanshe, various years).

central government expenditures (CGE), however, the increase was relatively slow. Thus, it appears that the ratio of defense expenditures to CGE was diminishing, from 15.13 percent in 1981 to 8.05 percent in 1993. The ratio increased again to 9.51 percent in 1994.

The available evidence suggests that China allocated more resources to economic development than to defense. The crowding-out of the defense industry was rather serious: The PLA began raising its own funds. In this context, it appears that the policy of conversion to civilian production constituted the main method by which the state lessened the financial pressure on defense enterprises.

The absolute amount of Chinese defense expenditures is a highly controversial matter. Figure 3.1 provides official data on defense expenditures from 1950 to 1994. However, the 55.07 billion yuan China announced for its defense budget in 1994 could not reflect actual defense spending: If we assume that real defense expenditure has remained constant since 1979 and calculate a new figure based on the average 1979–1994 gross domestic product (GDP) deflator of 2.78—thereby taking inflation into account—nominal defense expenditures in 1994 would be 61.35 billion yuan. At the same time, soldiers' income has increased, and equipment has been renewed and upgraded. Given the special status of the PLA, even under the government's shift in emphasis to economic development, we must conclude that total defense expenditures should be considerably more than 61.35 billion yuan.

We can establish wide upper and lower boundaries for actual Chinese defense expenditures using two separate estimation methods. If we assume that, following the 24.5 percent reduction in PLA personnel between 1985 and 1990, defense expenditures were reduced by a similar amount and that the ratio of defense expenditures to central government expenditures remained the same as it was in 1979, then the actual defense budget for 1994 would be 69.31 billion yuan. However, if defense expenditures grew at the same rate as GDP, they would be 183 billion yuan in 1994. If we consider these two amounts to be the upper and lower bounds of Chinese defense expenditures, then in 1994 the Chinese army somehow had to find between 14.24 and 127.93 billion yuan over and above the stated defense budget figures to compensate for the shortfall. This range is large, reflecting the uncertainty inherent in Chinese budget figures. Nevertheless, our calculations suggest that the shortfall is serious. This extra expenditure would have had to be supported by business revenue from defense and PLA enterprises.

Output value of defense conversion. There are various estimates of the output value of defense conversion. Arthur Ding calculates that it reached about 44 billion yuan in 1993.[6] Table 3.2 provides our own estimates, derived from industry yearbooks, of the value of civilian production from defense (non-PLA) industries. These incomplete time series demonstrate that civilian production increased very rapidly before 1985, with an average annual growth rate of over 20 percent. The growth rate slowed between 1986 and 1991 but remained higher than the rate of national economic growth. With rapid economic growth, the progress of defense conversion accelerated in 1992. In the two years after 1992, the growth rate of civilian production value by military enterprises reached 30 percent. Therefore, it is easy to see that the growth of military conversion to civilian production was affected both by policies described previously and the demand side; the speed of civilian economic growth also determines the progress of military conversion.

Nevertheless, there are no reliable statistics and estimates concerning civilian production, real estate sales, and tertiary industries directly operated by the military. According to an official announcement by the Chinese

Table 3.2 Civilian Production Value by Converted Military Enterprises (in billion yuan)

Year	Value of Civilian Production	Average Growth Rate (%)
1978	2.45	–
1980	3.67	22.4
1985	11.76	26.2
1987	14.95	12.8
1990	22.73	15.0
1991	24.39	7.3
1992	31.68	29.3
1993	41.45	30.8

Source: Calculated from Committee of Editors, *Zhongguo Gongye Nianjian* [China Industry Yearbook] (Beijing: Zhongguo Laodong Chubanshe, 1991, 1993, 1994).

Note: Numbers are based on constant 1990 prices.

government, the total value of civilian production by PLA factories was 2.71 billion yuan in 1987, making up two-thirds of the total production value in that year. The revenue from mining and the sales volume of tertiary industries run by the military were 4.5 billion yuan and 2.38 billion yuan, respectively. The total value contributed by the military was 9.59 billion, and the military gained 2.41 billion yuan in profit.[7]

Table 3.2 shows that the production value of converted non-PLA defense enterprises in 1987 was 14.95 billion yuan. If we consider the PLA's 9.59-billion-yuan production value, the total defense production value was 24.54 billion yuan in 1987.[8] If the ratio of profit to production value was 25 percent,[9] the increased budget for the military was 6.14 billion yuan, representing 29.2 percent of the 20.96-billion-yuan defense budget in that year. From this rough estimate, we can see that conversion policy did not significantly raise the defense budget; its contribution was about 30 percent. Despite the fact that the production value of converted enterprises has been very close to the defense budget in recent years, the benefit of conversion policy to defense expenditures should be determined by the profit of converted enterprises rather than the production value.[10] The production value of converted enterprises has more important linkage and employment effects for the national economy.

Efficiency. Before discussing the impact of converted enterprises on the macroeconomy, we must examine the efficiency of converted enterprises. If the efficiency of converted enterprises is lower than that of civilian enterprises, then conversion policy hurts resource allocation from the viewpoint of opportunity cost. Implementing conversion policy is then only useful in remedying the problems of idle capacity and underemployed labor. In this section, we will limit our comparison to converted industries of civilian enterprises in the machine industry.[11]

Table 3.3 compares the indices of converted enterprises and civilian mechanical enterprises. It shows that the performance of converted enterprises is lower than that of general civilian enterprises in the machine industry. In a market economy, inventory is usually considered investment. Inventory always increases in planned economies such as China's, however, because supply exceeds demand. This difference may be considered an indication of whether or not China's planning met the needs of the market. In 1991 and 1992, the inventory was 4–5 percent in civilian machine enterprises, whereas the inventory of converted enterprises was over 15 percent. This indicates that converted enterprises had a more severe production surplus than did civilian enterprises.

Regarding the ratio of profit to sales, civilian enterprises earned over 3 percent, whereas converted enterprises contributed only 0.69 percent and 1.74 percent in 1991 and 1992, respectively. The ratio of exports to domestic sales was over 25 percent for converted enterprises but was only 10

Table 3.3 Comparison of Converted Enterprises and Civilian Enterprises in the Machinery Industry

	1991		1992	
Item	Civilian Enterprises	Converted Enterprises	Civilian Enterprises	Converted Enterprises
Production value (billion yuan) (1)	540.49	23.94	737.31	32.38
Sales volume (billion yuan) (2)	512.72	20.24	693.19	26.98
Inventory change (billion yuan) (3)	27.77	3.70	44.13	5.90
Ratio of sales to production (%) (2/1)	94.86	84.56	96.02	83.33
Total profit (billion yuan) (4)	19.03	0.14	22.83	0.47
Ratio of profit after tax (%) (4/2)	3.71	0.69	3.29	1.74
Total export (billion US dollars) (5)	7.15	n.a.	13.25	1.27
Ratio of export (%) (5/2)	7.43	n.a.	10.52	25.81

Source: Calculated from Committee of Editors, *Zhongguo Gongye Nianjian* [China Industry Yearbook] (Beijing: Zhongguo Laodong Chubanshe, 1992, 1993).

Note: Numbers are based on 1995 prices.

percent in civilian enterprises. The products made by converted enterprises were apparently more exportable.

Calculating the impact. The foregoing analysis is a partial-equilibrium analysis based on certain assumptions. Next, we will use the analysis of industrial interdependence to explore the impact of defense conversion on China's macroeconomy. As several constraints limit available data, we must make stronger assumptions in studying these empirical results.

Most of the civilian products made by the military, such as home electronic appliances and motorcycles, are consumed as final goods. Some, such as automobiles and ships, are used for government consumption. Others are used for intermediate inputs, such as large-scale equipment for power plants and automotive and shipbuilding factories. In order to simplify the analysis, we consider all these products to be final-demand goods.

In the previous section we demonstrated that the performance of converted enterprises is rather different from the performance of general civilian production enterprises. Clearly, the input-output relationship between these two kinds of enterprises is also different. As there are no input-output tables for converted enterprises, we can only use general input-output tables and simulate converted enterprises' effects on the macroeconomy.

According to the input-output account, the balance equation is

$$AX + F = X + M \tag{3.1}$$

where A is an input coefficient matrix, X is a vector of production sectors, F

is a vector of final demand, M is a vector of import, and AX is a vector of intermediate demand. In this model, the import coefficient is

$$m = M/X. \tag{3.2}$$

Substituting $M = mX$ into equation (3.1), we get

$$X = (I - A + \hat{m})^{-1} F \tag{3.3}$$

Here m represents a diagonal matrix of input coefficients. The term $(I - A + m)$, a square matrix, stands for the level of interdependence. Its coefficient of the diagonal refers to the direct and indirect demand for that industry if the final demand increases one unit. The nondiagonal matrix represents the indirect demand for other industries. Its horizontal sum is the total demand for that industry, which is used to measure the interdependent relationship with other industries.

In addition to the changes in total output generated by interdependence, we are also concerned with the employment effect. Using the industrial inverse matrix $(I - A + \hat{m})^{-1}$, this section analyzes the direct and indirect employment effect created by final demand. The employment multiplier matrix, like the industrial inverse matrix, can be written as

$$E = \hat{I}\,(I - A + \hat{m})^{-1} \tag{3.4}$$

where $\hat{I}$ is a diagonal matrix and $\hat{I}$ stands for the ratio of output to labor.

Next, we will examine effects on the macroeconomy by using the input-output tables issued by China in 1987 and 1990, which covered thirty-three industrial sectors. We combine metal product, transportation equipment, electrical and electronic equipment, instrument and measuring instrument, and general machinery manufacturing into the general mechanical manufacturing category. In other words, we alter the input-output tables from thirty-three sectors to twenty-nine sectors to simulate the empirical results based on the foregoing procedure.[12]

According to the data in Table 3.2, the production value from converted enterprises was 14.95 billion yuan in 1987 based on 1990 constant prices and 10.76 billion yuan in nominal 1987 prices.[13] It increased to 22.73 billion yuan in 1990. Considering these two values as the final demand of civilian industry in a broad sense, and plugging these numbers into the input-output tables of 1987 and 1990, and using equation (3.3), the direct and indirect production values contributed by converted enterprises in 1987 and 1990 are 29.3 and 72.7 billion yuan; the multipliers are 2.7 and 3.2, respectively. Using equation (3.4), we estimate that converted enterprises created 1.72 and 2.67 million jobs in 1987 and 1990, respectively. Overall, defense conversion contributed less than 2 percent of total production value and total employment to China's economy (see Table 3.4).

Table 3.4 Effects of Defense Conversion on China's Economy

Item	1987	1990
Income effect (billion yuan)	29.3	72.7
Ratio to total production (%)	1.3	1.9
Employment effect (million persons)	1.72	2.67

Source: Calculated from Department of Balances of National Economy and Office of Input-Output, Survey of State Statistical Bureau [Guojia Tongjiju Guomin Jingji Pingheng Tongjisi, Guojia Tongjiju Tuoru Chanchu Bangongshi], eds., *Zhongguo Turoru Chanchu Biao* [Input-Output Table of China] (Beijing: Zhongguo Tongji Chubanshe, 1991, 1993).

Note: Numbers are based on 1995 prices.

Microanalysis of Converted Enterprises

Converted defense enterprises became involved in a wider scope of business and diversified operations as economic reforms deepened in China. In this shifting environment, how can the converted enterprises solve the problems of idle labor and capacity? How can they create new products and markets? What advantages do the converted enterprises have? What difficulties do they face? These questions must be answered from the micro side of enterprises. Here, we offer answers by examining the automotive industry as a case study.

The automotive industry. The demand for cars increased greatly with China's rapid economic growth after the economic reforms. However, automobiles in China are more expensive than those in other countries due to government protection of domestic automotive firms, which encouraged every province to invest in the automotive industry. Many defense enterprises have the appropriate technology for automobile production, and many became active in producing automobiles or automotive products. Automotive manufacturing depends on several upstream industries such as machinery, electronics, and materials.

Before we analyze the converted automotive enterprises, we must briefly illustrate the features of the automotive industry as a whole in China. In short, the characteristics of China's automotive industry can be summarized as follows:

- *China's automobile industry is fragmented.* There are 126 automobile manufacturing plants, 120 van assembly facilities, 500 special car assembly plants, and 3,000 car component factories. The total production of automobiles was 1.28 million units in 1993.[14]
- *The industry produces a wide variety of final products.* According to a survey conducted by the China Automotive Industry Corporation in 1993 that covered 78 key automobile factories, the product break-

down of automotive production was 58.6 percent trucks (3 percent heavy type, 26.8 percent medium type, 23.3 percent light type, 5.5 percent micro type), 22.2 percent vans (0.4 percent large to medium type, 1 percent medium type, 15 percent light type, 5.8 percent micro type), and 19.2 percent sedans.[15] Among the various types of automobiles, sedans had the highest growth rate—40 percent in 1993.[16]

- *Most automobile buyers are government organizations.* Private individual purchasers are rare. Auto purchases are one of the ways the government exercises economic control, and the volume of auto sales drops considerably during economic slowdowns. Nevertheless, the growth rate in automobile production seems rather stable. In 1981, the production of cars was only 175,000; it increased to 1.28 billion by 1993. The average annual growth rate was 18 percent between 1981 and 1993.[17]
- *China is weak in automotive research and development.* Chinese automobiles are old-fashioned and of low quality and tend to be models that other countries have stopped producing. As automotive manufacturing capacity is small and lacks economies of scale, automobile prices are usually two to three times higher than in other countries.

The central government has attempted to restrict local governments' efforts to build more automotive factories. The integration of small factories into large corporations was suggested in order to meet the requirements of economic scale. Nevertheless, the number of auto factories increased in the late 1980s. Table 3.5 indicates that the number of automobile and components factories increased at a higher rate in the defense sector than in the general civilian automotive sector. Indeed, in the early 1990s, defense automobile factories were still growing, while civilian automobile factories were scaling back. As for the number of enterprises, the share of automobile factories operated by the defense sector increased from 5.9 percent in 1987 to 10.5 percent in 1993.

The average size of defense automobile factories was larger than that of general automobile factories. It is evident that the ratio of defense auto production to total automobile production is higher than the ratio of the number of defense factories to the total number of automotive factories. However, defense auto enterprises also have a higher ratio of employees to output than do general auto factories. For instance, production by defense enterprises constituted 12.8 percent of total automobile production in 1993, but defense sector employees were 21.3 percent of total employees, demonstrating that defense factories have a considerable labor surplus. Such figures indicate that the labor productivity of converted enterprises is 55–90 percent lower than that of civilian enterprises in automobile manufacturing.

The performance of defense enterprises overall is worse than that of

Table 3.5 Production of Military Enterprises in the Automotive Industry

Item	1987 Industry	1987 Military Enterprises	1990 Industry	1990 Military Enterprises	1992 Industry	1992 Military Enterprises	1993 Industry	1993 Military Enterprises
Gross output value (10,000 yuan)	2,731,850	233,493 (8.5)	4,924,941	569,190 (11.6)	1,191,052	1,256,861 (10.6)	1,792,001	2,296,018 (12.8)
Net output value (10,000 yuan)	751,291	61,088 (8.1)	1,205,069	142,107 (11.8)	3,042,599	334,051 (11.0)	4,029,292	527,770 (13.1)
Added-value/gross output value (%)	27.5	26.1 (94.9)	24.5	25.0 (102.0)	25.5	26.6 (104.3)	24.0	23.0 (95.8)
Number of employees	1,348,267	167,810 (12.4)	1,565,332	261,051 (16.8)	1,809,512	348,682 (19.3)	1,932,575	412,366 (21.3)
Gross output value/number of employees (yuan/person)	20,262	13,914 (68.7)	31,462	21,804 (69.3)	65,821	36,046 (54.8)	92,726	55,679 (60.0)
Total pretax profits (yuan)	402,808	17,127 (4.3)	427,780	38,489 (9.0)	1,333,390	125,701 (9.4)	1,789,470	138,465 (7.7)
Pretax profits/total assets (%)	18.0	5.0 (62.5)	11.0	5.8 (52.7)	16.5	9.8 (5.9)	14.1	6.9 (4.9)
Number of enterprises	2,358	138 (5.9)	2,596	232 (8.9)	2,555	237 (9.3)	2,462	258 (10.5)

Source: Zhongguo Qiche Jishu Yanjiu Zhongxin [China Research Center for Automotive Technology], *Zhongguo Qiche Gongye Nianjian* [China Automotive Industry Yearbook] (Tianjin: Zhongguo Qiche Gongye Nianjian Biaojibu, 1988, 1990, 1993, 1994).

Notes: Numbers are based on 1995 prices. The numbers in parentheses represent military automotive enterprises as a percentage of the automotive industry as a whole.

civilian enterprises. Defense enterprises' ratio of pretax profit to production value was lower than that of civilian enterprises, which indicates that military enterprises have a lower operational ability. In addition, their lower ratio of pretax profit to total assets implies that military enterprises depend more on loans or other sources of capital than do civilian enterprises. If converted enterprises expand excessively, they will take away resources from civilian enterprises. Finally, defense enterprises had higher ratios of added value than did the automotive industry as a whole. This does not show that military enterprises engaged in higher technical production. In contrast, it implies that the division of labor between upstream and downstream defense enterprises is low.

Interestingly, China's aircraft factories have gradually moved into the automotive sector, producing large and small vans. For example, Shenyang Aircraft Factory, Harbin Aircraft Factory, Changhe Aircraft Factory, Shanxi Aircraft Factory, and Guizhou Air Cooperation are all involved in van production. The PLA's 7407 factory (Liaoning), 7424 factory (Shanghai), 7427 factory (Fujian), 6907 factory (Wuhan), 7435 factory (Hubei), 7319 factory (Hunan), 7429 factory (Hunan), 7436 factory (Sichuan), and 3402 factory (Shanxi) have also started to produce special cars, small vans, cross-country jeeps, and important automotive parts. In general, the defense sector remains outside the key automotive market—sedans and large to medium trucks—but it plays an important role in the production of micro-type trucks, light-type jeeps, and medium to small vans (used for minibuses).

Motorcycles. Military enterprises also play a crucial role in the motorcycle industry. According to Table 3.6, six defense enterprises are among the ten biggest factories. The first and second largest motorcycle factories belong to the ordnance system.

Table 3.6 The Big Ten Motorcycle Enterprises in China

Rank	Name of Enterprise	Production (ten thousand units)	Military Enterprise
1	Jialing Machinery Plant	59.00	yes
2	Construction Industrial Co.	58.85	yes
3	Chinese Light Motorcycle Co.	50.23	no
4	Shanghai Yichu Motorcycle Co.	23.62	no
5	Jincheng Mechanical Factory	19.01	yes
6	Chinese Southern Power Mechanical Co.	17.73	yes
7	Northern Yichu Motorcycle Factory	16.60	yes
8	Changchun Motorcycle Co.	10.78	no
9	Guangzhou Motor Co.	10.03	no
10	Yuhe Mechanical Factory	8.26	yes

Source: Zhongguo Qiche Jishu Yanjiu Zhongxin [China Research Center for Automotive Technology], *Zhongguo Qiche Gongye Nianjian* [China Automotive Industry Yearbook] (Tianjin: Zhongguo Qiche Gongye Nianjian Biaojibu, 1994).

The size of defense motorcycle enterprises is immense. Although their share of the total number of motorcycle producers is only 14.7 percent, defense factories made up over 50 percent of total output value in 1993. Moreover, defense motorcycle enterprises comprise 73.1 percent of the total employment in that industry, which shows that military enterprises involved in motorcycle manufacturing also have serious problems with surplus labor and low productivity. However, military enterprises have a higher ratio of added value, possibly because of the lower division of labor between upstream and downstream factories in this industry.

The ratio of pretax profit to production value was relatively low for military enterprises and was lower than that of general enterprises, because military enterprises employ labor more than capital. However, the profit per asset was higher than in the ordinary motorcycle industry (see Table 3.7).

Table 3.7 The Relative Performance of the Motorcycle Industry (1993)

Item	All Enterprises	Military Enterprises
Number of enterprises	75	11 (14.7)
Gross output value (billion yuan)	17.62	9.39 (53.3)
Net output value (billion yuan)	3.77	2.20 (58.2)
Added value/gross output value (%)	21.4	23.4
Number of employees (10,000 persons)	14.4	10.5 (73.1)
Gross output/number of employees (yuan)	122,361	89,429
Pretax profit/gross output value (%)	11.6	8.8
Pretax profit/total assets (%)	18.8	24.2

Source: Zhongguo Qiche Jishu Yanjiu Zhongxin [China Research Center for Automotive Technology], *Zhongguo Qiche Gongye Nianjian* [China Automotive Industry Yearbook] (Tianjin: Zhongguo Qiche Gongye Nianjian Biaojibu, 1994).

Note: The numbers in parentheses represent military enterprises as a percentage of the whole industry.

Future Trends

China's defense industry spreads across the Chinese economy, encompassing many different industrial sectors. After the economic reforms, the government reduced military production and emphasized production efficiency. Major parts of the defense sector—aviation, space, ordnance, nuclear—reorganized as corporations to solve the problems of surplus labor and excess capacity by transferring military production capacity to civilian production. Furthermore, profit maximization became the key objective for defense systems in order to underwrite military research and development. These were the original purposes of the implementation of China's defense conversion policy.

Currently, about 76 percent of all production by defense enterprises consists of civilian goods. The actual production value was 22.73 billion yuan in 1990. If we also consider the industrial interdependence of these military enterprises with the rest of China's industries by increasing the production figure about twofold, military enterprises created 72.7 billion yuan of production value. This constituted only 1.9 percent of the total production value and created employment for 2.67 million people. Military enterprises that produce civilian goods, then, have a substantial influence on many sectors, such as the automotive and motorcycle industries, but have no significant effect on the whole economy.

In the future, we believe conversion will remain a policy priority but that military enterprises will not have a significant influence on the ratio of civilian production to total production. Further improvement of the production efficiency of converted enterprises will be the focus of future conversion policy; converted military enterprises continue to face the problems of idle workers, excess assets, and low profit. In addition to improving the control, planning, and administration of converted enterprises, another solution is to clearly define the interrelationship between military and civilian production.

Basically, the state-owned enterprise still has not achieved the goal of independent management and responsibility for profits or losses. Transformation of these corporations into government-owned share corporations or further reorganization of existing ministerial corporations into corporation groups is the new direction of China's recent reforms. In order to effect these changes, China can pursue several alternatives:

- *Establishing separate subcorporations for military and civilian production under corporation groups.* Defense subcorporations would be completely supported by government-owned share corporations, whereas civilian subcorporations would be supported by private capital or joint ventures. Both would have independent financing.
- *Integrating all defense enterprises into a few enterprises specializing in military production.* Capital would be provided by the government, and the ministerial corporations would be transformed into corporation groups specializing in civilian production.
- *Not distinguishing between civilian and military production,* and executing a policy among enterprises of "civilian support of the military."

The third option is the current mode of operation; in addition to funding problems, however, the division of responsibilities between military and civilian production is not clear. Despite the fact that the second option would be suitable under present conditions in China, a shortage of funds and inefficient operation are obstacles to any new form of administration. Reorganizing defense production into a few specialized production institutions would be a difficult task, in which the initial problem would be finding the appropriate method of distributing benefits among the different ministerial corporations.

The first option is practicable: By retaining the same system of ministerial corporations (or corporation groups), it would be easier to clearly evaluate the real operational performance of both kinds of production. Based on individual advantages and disadvantages among subcorporations, the ministerial corporations could effectively implement planning and administration. Moreover, part of the profits from civilian subcorporations could fund defense operations.

Apart from the development of converted enterprises' systems discussed above, economic growth has also sparked another kind of defense conversion. Because defense expenditures have not grown as fast as have government expenditures, soldiers' income and military equipment have not been improved, and involvement of the army in production activities and use of the profits to support military needs have been the results. We estimate that such activities provided 9.59 billion yuan to the military in 1987, which represents about two-thirds of the total production value of converted enterprises. Although we lack direct and objective statistics, it is commonly believed that the growth rate of such PLA activities is probably higher than that of defense conversion industries.[18]

In recent years, the negative effects of directly involving the PLA in civilian production have attracted the attention of the Chinese government. Lower fighting capacity and military corruption are also suspected by the society as a whole. Some scholars propose that the government should root out these evils. In fact, China has taken some action to combat this problem, such as spending billions of yuan to take over a PLA coal mine in Shanxi, but more is needed. In addition to strictly strengthening military discipline, China must greatly raise its defense expenditures to satisfy the army's needs.

Given China's current fiscal position—the substantial increase in government deficits, the lack of funds for building infrastructure, and uneven income distribution—and the heavy demands of modernization and reform, it would be difficult for the Chinese government to considerably raise defense expenditures. The PLA needs to improve the lives of its troops and to upgrade its equipment—both expensive propositions. Accordingly, we can expect to see continued conversion and PLA commercial activities in support of these goals. These steps will also allow China to hide the scope of its actual defense spending, thereby avoiding the condemnation of neighboring countries. Whether or not this strategy—of which conversion is a part—will succeed is an open question.

Notes

1. Zhongguo Guowuyuan Xinwen Bangongshi [News Office, State Council of the PRC], "Zhongguo De Junbei Kongzhi Yu Chaijun" [Military Equipment

Control and Military Reduction in China], *Da Gong Bao* (Hong Kong), 17 November 1995, p. A7.

2. Ibid.

3. Tan Kai, "Jun Min Jiehe de Daolu Yue Zou Yue Kuan" [The Widening Road of Integrating Military and Civilian Production], *Zhongguo Gongye Nianjian* [China Industry Yearbook] (Beijing: Zhongguo Laodong Chubanshe, 1991), p. 130.

4. Ibid.

5. Guojia Jihua Weiyuanhui [National Planning Committee], "1992 Guofang Gongye Minpin Shengchan" [Civilian Production of Defense Industry in 1992], *Zhongguo Gongye Nianjian* (Beijing: Zhongguo Laodong Chubanshe, 1993), p. 132.

6. Arthur Shufang Ding, "Zhongguo De Guofang Jingji: 'Jun Zhuan Min' Dui Zhonggong Junshi Xiandaihua De Yingxiang" [Mainland China's Defense Economy: Impacts of Defense Conversion on Military Modernization in Mainland China], *Jingji Qianzhan* [Economic Outlook] (Taipei), May 1995, pp. 49–53.

7. Xu Guangyi, *Dangdai Zhongguo Jundui de Houchin Gongzuo* [The Rear Service of the Contemporary Chinese Army] (Beijing: Zhongguo Shehui Kexue Chubanshe, 1990), pp. 592–613.

8. The activities of the PLA are numerous. We do not include such items as the revenue from foreign trade companies set up by the General Logistics Department and military regions. In addition, the military receives some revenue from illegal businesses, such as automobile and weapons smuggling. On this, see Hu Angang, *Tiaozhan Zhongguo: Deng Hou Zhongnanhai Mianlin De Jiyu Yu Xuanze* [Opportunity and Challenge for China after Deng Xiaoping] (Taipei: Xinxinwen Chubanshe, 1995), pp. 183–201.

9. Suppose that PLA and non-PLA companies have the same ratio; 2.41/9.59 = 25.1 percent.

10. Most of the production value of converted enterprises is the transfer value of production factors, not added value created by converted enterprises.

11. The term "machine industry" (or "machine-building industry") has two definitions. In a broad sense, the machine industry includes metal products manufacturing, transportation equipment manufacturing, electrical and electronic equipment manufacturing, and instrument and measuring instrument manufacturing. In a narrow sense, it refers only to general machinery manufacturing. Here, we mean "machine industry" in the first, broader definition.

12. Regarding the concept and application of input-output tables, please refer to Victor Bulmer Thomas, *Input-output Analysis in Developing Countries* (New York: Wiley Press, 1982).

13. Calculated from the index of the machine-building industry, industrial products producer price index by category of commodities; Guojia Tongjiju [State Statistical Bureau], *Zhongguo Tongji Nianjian* [Statistical Yearbook of China] (Beijing: Zhongguo Tongji Chubanshe, various years).

14. Zhongguo Qiche Jishu Yanjiu Zhongxin [China Research Center for Automotive Technology], *Zhongguo Qiche Gongye Nianjian* [China Automotive Industry Yearbook] (Tianjin: Zhongguo Qiche Gongye Nianjian Biaojibu, 1994), pp. 47–62.

15. See China Research Center for Automotive Technology, *Zhongguo Qiche Gongye Nianjian,* pp. 47–62.

16. Calculated according to data from ibid.

17. Calculated according to data from China Research Center for Automotive Technology, *Zhongguo Qiche Gongye Nianjian,* various years.

18. Hu, *Tiaozhan Zhongguo,* pp. 183–201.

PART 2

The View from the PRC

4

Conversion and Restructuring of China's Defense Industry

Chai Benliang

With the reforms of the late 1970s, the policy of "military-civilian combination" was again raised in China. Since then, conversion has been implemented with remarkable results. Along with an outline of the development of that policy, major directives and achievements are presented below, followed by a thorough investigation of the business system. After a description of differences between the Chinese civilian and defense systems, specific steps toward a restructuring of the defense system are discussed.

Development, Directives, and Achievements

Preliminary attempts at a military-civilian combination policy were made immediately after the founding of the People's Republic of China (PRC). As early as 1952, the Central Ordnance Commission concluded that every defense factory should also produce civilian goods. Such a conclusion by a military institution reveals that severe shortages of civilian goods existed at that time; it also indicates that the perception of external threats was less severe in the early 1950s and that civilian restoration was viewed as imperative after many years of civil war.

As is demonstrated in Table 4.1, strong support for military-civilian combination came also from Mao Zedong and Zhu De in the late 1950s; the latter, one of China's famous military leaders, focused on a policy that may be described as "conversion to reconversion"—in other words, companies in the defense sector should be able to adjust their production according to varying threat perceptions.

Nevertheless, the overall environment during most of the period covered by Table 4.1 was not favorable to fostering that policy. Given the consequences of the Great Leap Forward, it is hardly surprising that few resources were available to overcome "civilian" problems in the late 1950s and early 1960s. In subsequent years, there were attempts to renew a

Table 4.1 Development Stages for a Defense Conversion Policy

Year	Initiator	Description
May 1952	Central Ordnance Commission in a report to the Party Central Committee (PCC)	Every military factory should also produce civilian goods
May 1956	Mao Zedong at a State Council meeting	More attention to civilian goods, which could be produced in peacetime by defense industries
April 1957	Zhu De in a report to the PCC	Combination of civilian and military production within the defense industries to make more civilian goods during peacetime and more military goods during wartime
1964	Zhou Enlai at the Third National People's Congress (NPC)	Reintroduction of "military and civilian combination" and peacetime and wartime production combination to help the national economy
1982	Deng Xiaoping's new military-civilian integration policy	Apart from fulfillment of military R&D and production in all defense industries, more emphasis on civilian production
1984	Deng Xiaoping at the Central Military Commission	Emphasize the technological superiority of the defense sector and that many possibilities with tremendous benefits and no harm exist to serve national construction needs

military-civilian combination policy; considering the difficult circumstances (from the Cultural Revolution to the successor struggle after Mao's death), however, it was only in 1982 that Deng Xiaoping was able to reinitiate this policy. As part of the overall reform of the 1980s, restructuring and conversion began inside the defense sector. Nonetheless, it must be clearly mentioned that, throughout these years, top priority was given to guaranteeing military product assignments.

The major directives and achievements during the first decade of the reform process—the 1980s—can be described as follows:

- *Work hard for the development of civilian products to meet the demands of society.* In order to develop the domestic market, defense industry departments mass-produced light-industry products and electrical appliances, such as cameras, bicycles, sewing machines, refrigerators, and motorcycles; thus, motorbike production by defense companies increased to more than 60 percent of the country's total.
- *Use the technical superiority of the defense industry to develop a variety of products with new, high-tech features.* Especially within the nuclear industry, many such products were developed and invented: minireactors, automatic fire-warning systems, nuclear and petrochemical valves, radiation-immune medical packages, nuclear

medical instruments, and nuclear radiation detection instruments. Aerospace industry units performed research and development (R&D) on new textile machinery, refrigeration equipment, and satellite communication systems as well as industrial robots and industrial controllers. Ordnance industry units developed coal-mining excavators, underwater electric pumps, oil rigs, sea beach oil trucks, and ore trucks. The defense industry system could produce fifteen thousand kinds of civilian products and many technical facilities for light industry, the textile industry, the foodstuffs industry, and the medical and health systems and helped reduce import quotas for high-tech equipment and technologies such as petrochemicals, chemical fertilizers, and chemical fibers, thus conserving significant amounts of hard currency for the country.[1]

- *Actively undertake key project construction, equipment procurement, and R&D for major items.* Companies within the defense sector supplied products, know-how, and personnel for major projects under the relevant Five-Year Plans. Examples may be found in the aerospace industry—in the manufacture of electron-positron colliders and in the production proportion of civilian aircraft, which grew constantly in the reform period. In addition, planes and helicopters contributed to forest-fire prevention, crop seeding, fertilizer spreading, weeding, insect elimination, disaster relief, aero-photography, wild-animal tracking, offshore oil-rig support, and shuttle passenger and cargo transportation. In the automobile sector, defense companies increased their share of output drastically, up to a share of 50 percent of the national total volume of minibus production, and up to a 60 percent share in total motorcycle production.[2]
- *Speed up the transfer of military technology to civilian usage, driving up national economic construction.* More than thirty-one thousand civilian technical questions were solved with the help of defense companies. For example, aero-turbine expansion refrigerating technology was applied to the recovery of natural gas from oil fields, salvaging millions of tons of gas and transforming "waste" into precious resources. After this directive was implemented, technology transfer contracts from the military to the civilian side accelerated. During the Seventh Five-Year Plan, the number of contracts in this field increased to almost thirty thousand.[3]

The results of all these guidelines are quite impressive: In the past fifteen years, the value of civilian goods produced by the defense industry has maintained a steady annual growth rate of over 20 percent. From 1979 to the end of 1994, the proportion of civilian production has increased from 8 percent to 80 percent of the total output value.[4]

Military and Civilian Production in the Chinese Defense, Science, Technology, and Industry System

After forty years of striving toward self-reliance and independence, China has established a complete and integrated DSTI (Defense, Science, Technology, and Industry) system that includes almost all major specialties and adequate facilities for research, development, test, and evaluation (RDT&E), and production. This system has made outstanding contributions to socialist construction and the safeguarding of national independence, security, and world peace and has developed both advanced strategic defense weapon systems and conventional weapon systems.

In 1991, the Chinese government published the Ten-Year Program and the Eighth Five-Year Plan for National Economic and Social Development, calling for adherence to the policy of "Reform and Opening" in order to accomplish greater achievements. The document requires the Chinese DSTI (CDSTI) to continue its conversion and the readjustment of the defense industrial structure, and every defense enterprise must have a "pillar" civilian product by the year 2000. Finding the appropriate way to achieve further reform, continue opening to the outside, and steadily integrate military and civilian production while safeguarding national security and independence is the historic task that the CDSTI must perform in the next decade and is the prerequisite for reform of the CDSTI in the 1990s.

The CDSTI is a large and complex social system, involving various elements of politics, the military, the economy, and science and technology. If one considers function, status, institutional structure, control, operational mechanism, and evaluation criteria, the forty-year development of the CDSTI may be divided into two stages: (1) from the founding of the People's Republic of China to the early 1980s, and (2) from the early 1980s to the present (see Table 4.2 for details).

Table 4.2 clearly demonstrates that, in the past fifteen years, the CDSTI has undergone system-wide reform. Tremendous changes have occurred, ranging from system function to evaluation criteria; nevertheless, these changes have not been implemented coherently. Therefore, system-wide reform has not yet brought about system effectiveness—which, if realized, will be greater than the aggregate effectiveness of all elements of the current system.

Although Chinese defense enterprises are, at their core, economic organizations producing military hardware, the CDSTI has a specific law for its development, as the nature of military products is distinct from that of civilian commodities. In addition, many differences between military and civilian development and operational processes exist (see Table 4.3).

The different commodity natures of military and civilian products dictate their different operational mechanisms. Military production in China is conducted through contracting and is subject to mandatory planning, where-

Table 4.2 CDSTI System Characteristics Comparison

Stage	System Function	System Status	System Structure	System Control	Economic Mechanism	Evaluation Criteria
Before 1980s	Mono-military production	Independent, closed system	Centralized pyramid	Administrative means only	Planned economy	Schedule, technical specifications
After 1980s	Military and civilian production	Relatively independent, open system	Flat, decentralized	Administrative, economic, legal, etc.	Socialist market economy	Schedule, specifications, cost, economic efficiency

Table 4.3 Comparison of Military and Civilian Commodities

Characteristics	Military Commodities	Civilian Commodities
Investment channel	Mono-government investment	Government and private investment
Resource allocation	Centralized planning	Planning and market
Supply and demand regulation	Planning, mandatory contract	Planning and market
Client	Military departments	Individuals
Market	Very limited	Fully developed
Price	Cost plus fixed profit	Market
Transaction form	Mandatory contract	Market exchange
Generation of demand	Military, noneconomic demand	Social, economic demand
Embodiment of demand	Government plan	Market demand
Program approval	By government	By private sector
Operation criteria	Technical performance	Technical and economic performance
Sale	According to plan	According to market
Maintenance	By the user	By the producer
Ownership of the enterprise	State	State and private

as civilian production is limited by the socialist market economy. Most of the CDSTI has already undergone economic conversion to some extent; therefore, there are generally two operations running simultaneously in one enterprise. At the same time, the state objective for enterprise reform is to give increasing authority to enterprises in civilian production management and operation, allowing them to become independent legal entities. As a result, incompatibility and a lack of coordination currently exist between military and civilian production sectors within each enterprise in the CDSTI. Other problems of the Chinese DSTI system are the two-tier price system (i.e., both market and official prices) and the improper profit sharing between workers in the military and civilian sectors of the same enterprise—a consequence of the fact that profits from civilian production are usually higher than those from military production.[5]

Overcoming these incompatibilities and conflicts will involve significant managerial efforts. The principal solution, however, must be to develop procedures to achieve synergy between the military and the civilian. In current Chinese circumstances, these procedures include a set of operational mechanisms and corresponding macro and micro controlling and coordinating tools, all of which are based on the law of the market economy. They will render the macro, mandatory planning of military production compatible with the operational mechanisms of the market economy and the law of value at the enterprise level.

Restructuring the Chinese Defense Industry

Over the past fifteen years, economic reform of the CDSTI has been administered at three levels. The reforms of the first tier deal with the transition from a planned economy to a socialist market economy. The major feature of such reforms in terms of the CDSTI is the transformation of industries run by the ministries of defense into general corporations. For instance, the Ministry of Nuclear Industry, a government administrative agency, was transformed into the China Nuclear Industry Corporation, an economic entity. The second-tier reforms occur at the level of the enterprise; in the past, enterprises were essentially budget-centered production organizations, but they have become profit-centered organizations dealing with production, marketing, and sales.

The reforms of the third tier involve changes in the operational mechanisms of enterprises. In the past, the operational mechanisms of the CDSTI were essentially budget- and plan-oriented. Managers were most concerned with carrying out production entirely according to the plan and within the budget, and paid little attention to cost and other economic measures of the enterprise's performance. All costs were paid by the government, and all profits were given to the government. The current profit-oriented mechanisms in China, however, require the enterprise to conduct its operations independently; operating under the socialist market economy, enterprises pursue maximum profit and enjoy great independence in their economic operations. All of the third-tier reforms are common to both civilian enterprises and defense enterprises in China. Nonetheless, some difference does exist in military production—it operates according to contracts negotiated between the client (the military) and the enterprise but is closely regulated by state acquisition strategies and plans.

Reform of Economic Operational Mechanisms

The reform of operational mechanisms in Chinese enterprises since 1978 may be divided into two stages:

- *Enterprise operational responsibility contracting system.* During the first stage—from 1978 to the late 1980s—reforms were implemented in many areas, but their primary purpose was focused on decentralization and preferential policies. Thus, the government granted some rights to enterprises and incentives to encourage managers to manage economic operations more effectively. During this stage, there was little reform governing the ownership and organizational structure of the enterprises.
- *Modern enterprise system.* The second stage has lasted from the late 1980s to the present. Reforms emphasize the operational mechanisms of enterprises, their organizational structure, and the relationship between the government and enterprises. In order to promote the construction of a socialist market economic system, several major issues concerning its establishment were discussed, and important decisions were made on 14 November 1993 at the Third Plenary Session of the Central Committee of the Communist Party of China (CPC) to speed up the steps of reform in order to complete the transition by the end of this century. Economic reforms of wider and deeper influence are therefore occurring.

Enterprise operational responsibility contracting system. The general operational mechanism of enterprises in the first stage is the enterprise operational responsibility contracting system. Under this system, the enterprise negotiates with the supervisory government agency on a contract. The contracts usually include the following aspects:

1. The enterprise is selected by the supervisory government agency for a contract between the two organizations.
2. The contract specifies the responsibility of the enterprise in its operations within certain periods of time: for instance, the base quota that

the enterprise should give to the government, the conditions that the government agency should provide in order to help the enterprise to carry out the contract, and so on.

3. The rights that both sides enjoy during the execution of the contract: for instance, the enterprise's rights in areas such as profit-sharing, personnel, and product selection, and so on.

Clearly, the enterprise operational responsibility contracting system deals chiefly with decentralization and incentive measures and does not involve such issues as ownership and organizational structure. The contracting system began to be implemented in 1982 and became the general working mode in 1987. At present, more than 90 percent of large and medium-sized state-owned enterprises (SOEs) have adopted this mechanism.[6] Defense enterprises have also implemented this contracting system.

The advantages of the contracting system as are follows:

1. Because the enterprise must give the government a fixed, contracted percentage of its income, the government receives an assured financial income from the enterprise.
2. After paying the contracted quota to the government, the enterprise has the right to dispose of its income and profits. Therefore, the managers and workers are encouraged to improve the economic performance of the enterprise.
3. Because the contracting system is carried out between different levels of the enterprise and everyone clearly understands his or her responsibility and economic reward, employees show more concern for production.
4. The contracting system is relatively simple to implement, as it involves only two organizations—the enterprise and the government agency in charge—and deals with only decentralization and some preferential policies.

Although the enterprise operational responsibility contracting system has many advantages, there are several flaws in the economic operation of the enterprises that cannot be overcome by the contracting system:

1. The overlap of government functions with those of enterprises. Under the contracting system, the enterprise remains under the supervision of the government agency in charge and is subordinated to it; the enterprise continues to depend heavily on the government in economic operations. Even in contracting, the government agency is not only the negotiator but also the approver of the contract. Thus, the government is considerably involved in production, marketing, sales operations, and capital investment in the enterprises. The func-

tions of the government in the management of enterprises are in many aspects identical to the functions of the manager of the enterprise.

2. The proportion of deficit-ridden enterprises in total state enterprises cannot be effectively reduced by the contracting system. The articles in the contract stipulate that when an enterprise fails to fulfill the agreed-upon targets, the losses must be covered by the enterprise itself. In reality, however, there are no other sources than the government from which the enterprise can obtain money to cover the losses. Under these circumstances, enterprises commonly ask the government agency to revise the relevant article to lessen the fixed quota.
3. The contracting system cannot overcome nearsighted economic behavior. The contracting term is usually relatively short, approximately two to five years. Hence, the enterprise manager's major task is to fulfill the targets assigned by the contract. Under such circumstances, long-term economic effectiveness is frequently neglected. In some cases, enterprise managers use all possible means and measures to fulfill the contract target, even at the expense of the long-term interests of the enterprise.
4. Better allocation of production elements—such as equipment, capital, labor, factory buildings—cannot be realized by the contracting system. The subordination of the enterprise to one government department prevents the enterprise from cooperating with other organizations subordinated to other government departments.

The enterprise operational responsibility contracting system has also been implemented under the CDSTI. Defense enterprises make contracts with general corporations—industries formerly run by ministries of defense—which, in turn, make contracts with the State Council. A portion of the profits set by the contract flows from the enterprise to the general corporation to the State Council. The rest of the profit will be disposed of by the enterprise itself. The contracts between the enterprise and the general corporation, or between the general corporation and the State Council, cover both military and civilian production. Because military production has been reduced sharply and its productivity is much lower than civilian production, the total profit from defense enterprises in general is lower than that from civilian enterprises. All the profits or losses, from both military and civilian production, are handled among the enterprise, corporation, and State Council; there is no connection with the military. The procurement contract between the armed services and the enterprise or the general corporation (in the case of major weapon systems procurement) deals only with the price, the technical specifications, and the delivery schedule. The military is not involved with the economic performance of the enterprise.

In sum, the enterprise operational responsibility contracting system was a temporary economic operational mechanism that worked under the macroeconomic circumstances of a planned economy and the market regulations put forward in 1982. After 1992, when the Party and the government decided at the 14th Central Committee of the CPC to make the socialist market economy the objective of economic reform in China, this system was inevitably replaced by another—the modern enterprise system.

Modern enterprise system. In the spring of 1992, Deng Xiaoping made several speeches during his investigative tour of southern China. In these speeches, Deng clearly pointed out that the planned economy is not equal to socialism because planning also exists in capitalism and that the market economy is not equal to capitalism because market forces also exist in socialism. Planning and market forces are both economic means. Deng's assertion not only was an important breakthrough in socialist economic theory but also eliminated many of the ideological constraints that had been imposed upon the reforms in China's economic system. In order to drive state-owned enterprises toward the market and to increase their economic vitality, the State Council issued the "Regulation of the Transformation of Operational Mechanisms of the State-Owned Enterprises" on 23 July 1992. The regulation attempted to transform enterprises into legal economic entities that bear civil rights and civil liabilities. On 12 October 1992, a decision was made by the 14th Central Committee of the CPC that the target of economic reform in China is the establishment and perfection of a socialist market economic system before the end of this century.

In order to speed up reform, opening to the outside, and socialist modernization in China, a "Resolution on Several Major Issues in Establishing the Socialist Market Economic System" was passed on 14 November 1993 at the Third Plenary Session of the 14th Central Committee of the CPC. This document states that the goal of economic reform is establishing a modern enterprise system and that the transformation of state enterprises into corporations is a beneficial way to achieve this end—standardized corporations can effectively realize the separation between the proprietary right of the investor and the property rights of the enterprise as a legal entity, as well as the distinction between the government function and the enterprise function in economic operation. More importantly, the state would bear only limited liability to the enterprise rather than the unlimited liability of the past. On 29 December 1993, the Corporate Law was passed by the National People's Congress in order to standardize the organization and behavior of corporations; protect the lawful interests of corporations, stockholders, and creditors; maintain the social economic order; and promote the development of the socialist market economy.

Since 1992, certain defense enterprises have been transformed into joint stock corporations. All of these enterprises are military-civilian combination enterprises; by law, 100 percent military production enterprises are not allowed to become joint stock enterprises.

Chinese defense enterprises, including those selected to be joint stock enterprises, have several features in common: (1) the coexistence of military and civilian production, which are managed by different operational mechanisms; (2) the coexistence of manufacturing units and support services, such as department stores, schools, kindergartens, and other services; (3) the coexistence of professional productive equipment and facilities and general purpose equipment and facilities, such as power stations and water supply systems. In the process of organizing joint stock corporations, these features must be taken into account.

Joint stock corporations in the CDSTI currently come in five fundamental forms, as demonstrated by the following examples:

1. *The Guangzhou Shipbuilding International Joint Stock Corporation.* The corporation was established in 1992: The productive operational property of the enterprise was transformed into a joint stock corporation, and the nonproductive part was separated and transformed into another economic entity. Because it is an international corporation, the auditing and evaluation work are done by both Chinese and foreign lawyers and accountants. The state stocks of the corporation are held by the China National Shipbuilding Corporation. The stocks enter stock markets both in China and abroad. State stocks account for 42.61 percent of the total; stocks outside China account for 31.82 percent, while stocks within China account for 25.57 percent.[7]
2. *The Changling Group Corporation* was created from the Changling Machinery Factory. Under this corporation, there are two organizations: the Changling Joint Stock Corporation, which produces only civilian products, and the military production organization. The Changling Group Corporation is therefore the holding corporation, and its stocks are issued within China.
3. *The Changhong Joint Stock Corporation* originally was the civilian production part of the Changhong Machinery Factory. The Changhong Machinery Factory is now the holding corporation, with stocks issued within China.
4. The original *Jialing Machinery Factory* has been divided into three parts. The civilian production part transformed into a joint stock corporation, the military production part remains a defense factory affiliated with the Ordnance Industry General Corporation, and the nonproductive part became another economic entity. The Ordnance Industry General Corporation is the holding corporation; its stocks are issued within China.
5. *The China Aviation Technology Shenzheng Corporation* has selected three subordinate enterprises to be transformed into joint stock cor-

porations. The Shenzheng Corporation is the holding corporation. One of the joint stock corporations, the Fiyta (Group) Joint Stock Corporation, has issued its stocks both within and outside of China.

The Influences of the Modern Enterprise System upon Defense Enterprises

Accompanying the development of the socialist market economy in China, profit-oriented principles will continue to penetrate deeper into the CDSTI system. Although the reform measures adopted in the past fifteen years alone cannot satisfy this new demand, the modern enterprise system—in conformity with the market economy—has no doubt effected far-reaching changes in the CDSTI, some of which have strategic significance. These changes may be summarized as follows:

1. *The investment channel has been broadened.* Joint stock corporations can collect capital for civilian production from the whole society and even from abroad. For instance, the Guangzhou Shipbuilding International Joint Stock Corporation has collected one billion yuan through the issuing of stocks in Shanghai and Hong Kong.
2. *Productive capability, both in quality and quantity, has improved.* The competition in civilian products is usually stronger than in military products. The joint stock form can help enterprises accumulate more capital to improve their manufacturing and developing capability through upgrades of machinery and facilities.
3. *The modern enterprise system has transformed the organizational structure and the managerial system.* An ordinary defense enterprise is led by one person, the head of the enterprise, who is responsible to the government agency in charge. In a joint stock corporation, on the other hand, the supreme authority is the stockholders; the board of directors is the operational decisionmaking organization and is responsible to the stockholders. Day-to-day management of the corporation's operations is conducted by the general manager. In addition, a supervising committee, which is also responsible to the stockholders, performs the auditing and supervising functions. Its members are drawn from representatives of the stockholders and the workers' union.
4. In an ordinary defense enterprise, the investor is the state, and the liabilities of the state to the enterprise are unlimited; when an enterprise is debt-ridden or bankrupt, all the remedial responsibilities rest with the government. *In the modern enterprise system, however, corporate law has clearly prescribed the liabilities of the investors and the operator of the enterprise.*

Restructuring in the 1990s

The restructuring of the CDSTI in the 1990s continues to have difficulties. In comparison to conversion in the West, restructuring of China's defense industries involves more aspects—the readjustment of industrial and product structures of the CDSTI as well as the operational mechanisms of civilian and defense production in defense enterprises. Meanwhile, a historic transformation of the economic system into a socialist market economy continues on both the macro and micro levels. Therefore, restructuring of the CDSTI not only must overcome many inherent problems of conversion but also is occurring in a fluid and dynamic environment unique to China, making the restructuring process more complicated and difficult than in the West. Nevertheless, the fifteen-year experience of conversion practices in China has brought understanding and enlightenment to further restructuring in this century. The main tasks of the future include the following:

1. *Continuing the readjustment of the industrial structure of the CDSTI according to state technology and industry policies.* The general guideline for readjustment is to develop civilian products in energy, transportation, and communication, which are bottleneck industries. Specific adjustments to be encouraged in defense industries include the following:
   - the nuclear industry: civil nuclear power stations, isotopic and radiation technology, nuclear instrumentation, and other relevant technologies;
   - the ordnance industry: machinery, chemical engineering, optical-electrical technologies, and various types of cars and vehicles;
   - the space industry: civilian satellites and satellite application technologies and products;
   - the electronics industry: information, computing, and automation technologies; and
   - the aviation industry: civil aircraft and nonaircraft products.
2. *Continuing reform of operational mechanisms in defense enterprises.* Emphasis will be placed upon the improvement of the enterprise operational responsibility contracting system and the timely evaluation of the experiences of joint stock corporations in the CDSTI, to ascertain whether the joint stock form could be the general reform mode in the CDSTI and how this reform mode might be implemented throughout the defense industry.
3. *Improving infrastructure that is indispensable for socialist market economic operations, and promoting the managerial ability of defense enterprises to enter into market economic operations.* For instance, most of the capital for civilian production in defense enter-

prises will come from the capital market instead of the state, as in the past. Enterprise managers must learn how to obtain loans and raise funds, and the government must learn how to establish and manage capital markets.

4. *Protecting military production in a military-civilian combination enterprise operating under the socialist market economy.* Profits from military production tend to be lower, requirements higher, and quantities less than in civilian production. Motivating enterprises and maintaining an adequate industrial base for military production under a tight defense budget are therefore questions that have become important under the CDSTI.

Notes

1. *Zhongguo Jungongbao* [Chinese Defense Industry], 30 September 1994, p. 1.

2. Ibid.

3. Cao Shixin, ed., *Zhongguo Junzhuanmin* [Conversion in China] (Beijing: Zhong Jingji Chubanshe, 1994), p. 15.

4. Information from the China Defense, Science, and Technology Information Center, Beijing.

5. The prices for military equipment are at present fixed by the government under strict control, whereas the prices for civilian products are determined by market demand and are therefore flexible. Consequently, the potential for profits is larger in civilian business.

6. Zhongguo Qiye Guanli Peixun Zhongxin Bian [China Enterprise Management Training Center], *Qiye Zhuanbian Jingying Jizhi Lilun Yu Shijian* [Transforming the Enterprise Operational Mechanism: Theories and Practice] (Beijing: Zhongguo Renmin Daxue Chubanshe, 1992), p. 90.

7. *Zhongguo Gaoxin Jishu Qiye Pingjia* [Review of China's High-Technology Enterprises], no. 3 (1995), p. 22.

5

Defense Conversion in the Chinese Press

Ka Po Ng

On 4 July 1995, *China Defense Industry* (*Zhongguo Jungong Bao*) celebrated its tenth anniversary with the publication of its 541st issue. It first appeared as a virtually unknown, restricted-circulation newspaper with the name of *Information Bulletin on the Military Technology Conversion for Civilian Use* (*Jungong Jishu Zhuan Minyong Xinxi Bao*).[1] It was born less than a month after the Central Military Commission (CMC), in an enlarged meeting on 4 June 1985, abandoned the doctrine of preparing the country for fighting an "early, all-out nuclear war" and ordered a demobilization of one million soldiers from the People's Liberation Army (PLA). China thus entered the stage of "peacetime construction," and the defense industries[2] embarked upon their "second buildup" (*dierci chuanye*).[3] In the same year, the China Association for Peaceful Use of Military-Industrial Technology (CAPUMIT) was formed to promote defense conversion. In retrospect, the past decade has witnessed changes in the international security environment, the evolution of China's defense industrial policies, personnel reshuffling, and ups and downs for defense industries. In adjusting to these factors, the newspaper has also found itself changing in almost every aspect, ranging from its mission and focus to its layout, content, and even title.

It was not until 1 July 1990, when the newspaper was renamed *China Military Civilian Production Integration* (*Zhongguo Junzhuanmin Bao*) and restructured accordingly that it became a domestic instrument to promote party and government defense industrial reforms, an open forum for defense factory managers to express their opinions, and, externally, a medium for the exchange of information between China and the world.[4] It no longer constituted a one-way channel for advising policymakers. Thus, alterations in the newspaper have been important indicators of the extent to which the Chinese have changed their attitudes and policies toward defense conversion.

The objective of this chapter is to examine Chinese defense conversion through the eyes of Chinese civilian administrators—the military and defense industrial plant managers. We will explore (1) the mission and structure of *China Defense Industry* and its evolution since 1991; (2) through reports in that newspaper, the Chinese concept of defense conversion; (3) problems perceived and solutions sought regarding defense industries; and (4) the impact on the military.

The Newspaper

Since the changes in 1990, the newspaper has been published in four pages, two times a week (Tuesday and Friday). It is under the administration of the Commission on Science, Technology, and Industry for National Defense (COSTIND) and receives guidance from the "Three Commissions" (*sanwei*), that is, COSTIND, the State Planning Commission (SPC), and the State Science and Technology Commission (SSTC).[5] According to the latest available information, the director of the newspaper is Xu Junli, assisted by deputy director Bi Xiaoqing and editor-in-chief Zhu Ercai.[6]

Signs of Openness: Masthead Changes

The newspaper has evolved in accordance with the direction and pace of defense conversion in China. Its reform in 1990 indicated a recognition of the necessity for a further opening of the defense industrial sector.[7] The most obvious, yet often neglected, sign of this opening was the newspaper's English subtitle: *China Military Civilian Production Integration.* Chinese publications rarely bear official English titles unless they are intended for international attention. At the same time, the newspaper became available for overseas subscription. By late 1989, the Three Commissions had urged the integration of defense industries with the national economy in their review of the past decade of defense conversion, and their subsequent "Recommendations for Further Civil-Military Integration" were endorsed by the CMC.[8] In 1990, China led a delegation to Moscow for the first international conference on defense conversion sponsored by the United Nations.[9] In turn, China hosted another meeting in Beijing in 1991, marking a new openness regarding the Chinese defense sector.

Months later, the English subtitle was replaced by Deng Xiaoping's "16-Character Slogan" on defense industry reform: civil-military integration (*junmin jiehe*), integration of peacetime construction with war preparedness (*pingzhan jiehe*), the priority of military production (*junpin youxian*), and support for the military from civilian production (*yimin yangjun*). In October 1993, however, an English subtitle—*China Defense Conversion*—returned, replacing Deng's slogan against a background of accelerated opening. Among the main events of that period were an international conference and an exhibition on defense conversion in Hong Kong in July 1993. These were masterminded by CAPUMIT and were claimed to be great successes by the Chinese defense industrial authorities.[10]

The newspaper arrived at its present condition in July 1994. Communist Party chief Jiang Zemin provided the calligraphy for a new Chinese title—*Zhongguo Jungong Bao*—and the English subtitle was again changed, to the current *China Defense Industry.* Moreover, uncommon for the Chinese press, a logo for the newspaper itself was introduced and was attached with explanations. As the editorial stated: "*China Defense Industry* . . . takes the perspectives of the defense industries, serves the whole country and actively promotes international exchange."[11]

The Mission, Layout, and Content

From its earliest days, the newspaper clearly depicted its mission as to guide and serve defense industries. Nevertheless, it tended toward one-way communication: It was largely a mouthpiece for the official policy line and for the government's "right" direction of reform. The belief that the most serious problem facing defense industries was a lack of information and guidance was reflected in the columns of the newspaper, which emphasized theoretical studies and polemics.[12] Defense conversion and the shift of development strategy from inward to outward looking were in their "preliminary stages," and the Chinese style was to use propaganda to pave the way for policy implementation.

Although senior officials were slow to recognize the deficiency of unidirectional communication, the editorial content of *China Defense Conversion* started to address this problem in 1993 when its role as a bridge began to be emphasized.[13] Apart from "digesting the directives from above in order to guide the sectors below," the newspaper had to "convey the situations and feelings at the lower levels upward." In accordance with this change, there were new columns such as "The Information Bridge" and "Letters to the Problem Solver." Moreover, Deng Xiaoping's inspection trip to southern China in 1992 accelerated reform not only in the economic sphere but also in the newspaper. Although *China Defense Conversion* had to organize special features to study the patriarch's speeches, it attempted to avoid hollow essays. At the same time, in addition to the tasks of guidance and service, it gained a new mission—being "practical" (*shijian xing*). A new economics page was devoted to discussions of the market, management, and production, which were closer to real life and more applicable to the concerns of plant directors.[14]

China Defense Conversion was designed mainly for defense plant directors, with "short, new, lively and in-depth" stories. Beginning in 1994, the newspaper reformed its third page and began to carry news synopses and columns on the operations of defense industries. With the change in title to

China Defense Industry in July of that year, the paper became more down-to-earth and worked harder to meet the needs of frontline defense industrial managers. Although theoretical studies continued, a new column, "Emergency University" (*yingji daxue*), was introduced to provide information on and offer possible solutions to critical problems. In 1995, the newspaper carried primarily easy-reading short stories and reports of no more than three hundred words. Even the column on political work and Party life, which had been full of rhetoric and theory, began to focus on personal experiences and case studies.[15] These developments indicated a gradual shift from overemphasis on vertical communication to a more balanced approach that considered the horizontal dimension as well.

Again, the changes in the newspaper over the years reflected the course of China's defense industrial reform—from the theoretical to the practical and from policy input to implementation output. In fact, the Chinese concept of conversion, the problems posed, and the solutions sought have been changing. Given the Chinese policy of "positive reporting," we may not be able to observe the difficulties that plague factory bosses in detail; however, through *China Defense Industry,* we may see existing trends, if not their true extent.[16]

Chinese Defense Conversion: Transformation Or Diversification?

Although Western literature defines defense conversion as the transformation of economic resources from military to civilian purposes, the policies of countries engaged in such exercises often show that they understand defense conversion in different terms and under their own rationales. This is particularly true for China.[17] Before we investigate the Chinese concept of conversion, however, we must ask two basic questions: Are there any particularly Chinese concepts of conversion, and if so, how many?

Junzhuanmin

Speaking at the 1990 Moscow conference, Huai Guomo (then major-general), deputy director of COSTIND and leader of the Chinese delegation, linked the Chinese conversion program to the international disarmament movement and focused on the one-way transfer of military technology to civilian use. He followed the same line in his speeches at later conferences on conversion, as did other Chinese delegates and news reports in *China Defense Industry* on similar occasions.[18] The term that has been used, "military-to-civilian conversion" (*junzhuanmin*), is noteworthy. According to Jin Zhude, chairman of CAPUMIT until 1995, the definition of this concept was confusing in China until the 1993 international conference in Hong Kong, during which the concept was defined: "Despite the different approaches and forms of military-to-civilian conversion, their programs are by and large the same, which include the conversion of defense production capacity to civilian purposes, peaceful use of military technology, the transfer of military equipment and facilities to civilian uses and the job placement of military personnel in the civilian society."[19]

Nevertheless, in the same review article of China's conversion efforts, Jin Zhude—in his capacity as the head of the Defense Conversion Liaison Group under the Three Commissions—pointed out that the phrase "military-to-civilian conversion" in itself did not provide a complete picture. It captured only part of the meaning of Deng's "16-Character Slogan" (which had been the official line of defense industrial reform) because it was different from "support the military with conversion" (*baojun zhuanmin*). He clarified the concept of military-to-civilian conversion in the context of the Chinese defense conversion policy:

> This 16-character principle, to be more concise, includes both "guarantee of military interests" [*baojun*] and "conversion to civilian use" [*zhuanmin*]; the latter is the so-called military-to-civilian conversion [*junzhuanmin*]. Thus, military-to-civilian conversion is not a directive but is a facet of the civil-military integration directive. This should answer the queries of those who have doubts about the term, military-to-civilian conversion.[20]

Baojun Zhuanmin

Certainly, Jin Zhude was not alone in this sentiment. As seen from the article in *China Defense Industry,* he summarized the official view in a succinct and precise way. Nonetheless, several observations may be drawn from Jin Zhude's elaboration of the Chinese concept of defense conversion. First, China has adopted a broad concept of conversion, which concerns the defense industrial base as well as all resources (human and hardware) that have been devoted to the military. The production and business activities (*shengchan jingying*) of PLA units are also regarded as part of the conversion program.[21] Indeed, army business units began to thrive in the late 1980s when Chinese annual military spending reached historic lows. According to 1993 official figures, these units made an annual profit of 5 billion yuan. Although the evidence is incomplete, it is believed that these earnings have partially helped to compensate for the insufficient budget.[22]

Second, as many recent studies point out, Chinese defense conversion efforts cannot be understood without taking details of its development strategy of civil-military integration into account.[23] The key to understanding centers on the "16-Character Slogan." In accordance with the principle of "priority to military products," the sine qua non of military-to-civilian conversion is to meet national defense requirements and, in fact, the purpose of injecting military technology into products of commercial value is to serve the ultimate objective of using civilian production to support the military. Production of civilian goods thus becomes a dynamic way to conserve military production capacity by keeping factories operating and technical workers in place. Quite often, when local authorities and defense plants report successes, they make sure to emphasize their fulfillment of the "guarantee of military interests" mission.[24] This emphasis may appear superfluous but is vital to the army; as demonstrated below, the true process of conversion seriously undermines military production capacity.

Third, the process of conversion is not only reversible, as Mel Gurtov observes,[25] but also works in two directions—from the military to the civilian sector and vice versa.[26] This is a matter of both principle (civil-military integration and peace-war construction integration) and necessity. As the processes of conversion and defense industrial reform unfolded, defense industries became aware of their technological deficiencies and recognized that they no longer enjoyed an unquestionable technological edge over their civilian counterparts. The application of civilian scientific research, technology, and equipment to military production therefore appeared necessary. The message of the two-directional flow of resources was unambiguously stated by the minister of COSTIND, General Ding Henggao, at a national science and technology conference in mid-1995. Dual-use technology was a top priority in the national defense science and technology development plan, and defense plants were urged to learn about and absorb civilian technology. This conforms to the CMC's strategy of "building a qualitatively superior army" and "strengthening the army technologically."[27]

Nevertheless, the principle of integrating peacetime construction with war preparedness deserves further attention. It involves more than the retention of military production capacity and strategic materiél reserve; it means the establishment of an efficient mobilization system. The defense industrial base is required to have a rapid response capability, and in the case of war, factories must be able to return to military production on short notice.[28]

Diversification Versus Conversion

In sum, economics and national defense modernization go hand in hand in the Chinese concept of defense conversion. Although official policy holds that conversion is a long-term policy (not an expedient), it does not point to a permanent transfer of military resources to the civilian sector or a transformation of defense research and development facilities and the industrial base. Are the Chinese then practicing diversification, rather than conversion, when their defense factories produce civilian goods? The answer depends on how one interprets conversion—the concept has undergone changes not only in China but also in Western literature.

Rather than establishing a unique approach, China has taken advantage of positive and negative Western experiences. Civil-military integration, which aims at the preservation of military production capacity in an economically viable way, is nothing new in the West, and the development of dual-use technologies and the application of civilian technological results to military hardware are not Chinese initiatives.[29] If there is anything especially "Chinese," it is not in the concept of conversion but rather in its implementation. China's course of political and national defense development since 1979 has brought special problems and has thus required special ways of tackling them. These have been evolving even as the process of conversion proceeds. We will see, through the lens of *China Defense Industry,* the problems perceived by the political and military leaders and the defense plant managers and how they plan to solve them. We will then look at the impact of conversion on the military as voiced by the latter in the press.

Perceived Problems and Solutions

When Deng Xiaoping launched his Four Modernizations program, he knew that the redundant production capacity of the defense industrial base and its irrational structure would be obstacles. But the patriarch could offer no more than his "16-Character Slogan" and piecemeal, pilot reform plans. It was not until a shift in grand strategy from war preparation to peacetime construction was announced in 1985 that Chinese conversion reached full swing. The late 1980s witnessed a frenzy among defense industries to branch out and develop whatever civilian products they could.[30]

The ensuing rapid economic growth was accompanied by serious inflation, however, and came crashing to a halt in 1989 amid political and economic conservatism. The complaints covered in *China Defense Conversion* at the turn of the decade focused mainly on macroeconomic issues. Only after Deng's push for accelerated economic reform did managers and directors show an awareness of the backward operational mechanisms of factories. There were arduous discussions on how to "Smash the Three Irons" (*za santie*): "iron chairs" (secure job positions), "iron wages" (secure wages), and "iron rice bowls" (secure jobs). In 1993, the focus swung to management problems, especially personnel issues. One year later, the newspaper's discussion of the problems became not only more diverse but also more sophisticated. In 1995, the military became more vocal in expressing its concerns about the neglect of military production. Thus, the problems of defense conversion raised in *China Defense Industry* have varied at different stages of development. At the same time, it is interesting that one item has never been omitted from the list of concerns: funding.

The Search for Macroeconomic Support

As a result of the macroeconomic adjustments that followed the serious inflation in 1988 and 1989, as well as the return to political conservatism after the military suppression of 1989, military plants faced a stringent fiscal policy, a hostile capital-raising environment, and, worse still, a weak market. When the Seventh Five-Year Plan ended and defense industries entered the 1990s, the complaints heard most often in *China Defense Conversion* were over macroeconomic support and financing, which either were not synchronized with the pace of conversion or were just not forthcoming. The central government, on the other hand, tended to attribute poor economic performance to the thinking (*sixiang*) of the factory bosses and their poor adaptation to the market mechanism as well as the lack of information.

The year 1991 saw several stories about insufficient macroeconomic support. According to the director of the Hunan Office in charge of Science, Technology, and Industry for National Defense, the plight of defense factories was due to unequal competition with civilian counterparts. Among the most prominent problems were irrational capital flow controls, lack of financial inputs, and short supply of raw materials. A plant director expressed dissatisfaction with a state bank that refused to implement government-approved loans.[31] Similar problems were not difficult to find in the newspaper.[32] They likely indicated that defense factories had not yet awakened from their dreams of the good old days during which they were provided for and protected by the government.

Although it was widely accepted that defense industries had difficulty surviving in the market at that time, there were disagreements over the causes of and solutions to the problem. Given a weak domestic market, calls for opening to the world became louder. Jin Zhude's article (on the front page) made an interesting contrast to the previously mentioned stories (on page three). The article, entitled "A New Stage in Military-to-Civilian Conversion—Advance to the International Market," urged factories to expand exports, invite foreign investment, and even enter into joint ventures so that they could gain a bigger market for their civilian goods and solve the problem of insufficient funding. Of course, success would depend upon improvements in product quality and reforms in the product mix.[33] Nevertheless, looking outward was not enough; others found poor information about the market to be another major weakness in defense industries, causing their slow responses to market signals.[34]

Complaints about the lack of macroeconomic support were not groundless. Defense industries were on their own; they had no choice but to "explore their potential" (*waqian*) and carry forward their tradition of "hard struggle" (*jianku fendou*) in the days of stringency.[35] Later events, however, revealed that the reputed "potential" and "technological superiority" of the factories were empty slogans. Despite the central government's decree in 1989 to integrate the defense industrial base into the civilian economy, defense factories were still not welcomed by local economic authorities and individual industries. The provincial governments either reluctantly urged the "faithful implementation" of policy or slowly provided detailed policy instructions. Integration therefore remained at a very preliminary stage.[36]

Apart from the focus on the unfavorable macroeconomic environment for defense conversion, discussions were also initiated in the newspaper about reforms in the operations of defense plants. When Deng emphasized in 1992 the spirit of "rush" and "breaking new ground," these debates came to a climax. Nevertheless, the reforms turned out to be conceptually over-simplified and, in policy terms, overshadowed by the zealous "Smashing the Three Irons" movement.

The Three Irons and Personnel Management

At the beginning of 1992, *China Defense Conversion* was set for exchanges about restructuring the inefficient operating mechanism of defense industry factories. In April, less than two months after Deng Xiaoping's inspection trips, it published its first stories on Smashing the Three Irons in military research and development units and defense factories. As presented by these reports, the "smashes" mainly took the form of promoting the factory responsibility system and introducing a contract system for employment, a differentiated wage system, and a promotion/demotion system based on job performance. This was fueled by criticism of formalism (*xingshi zhuzi*).[37] The inevitable consequences, however, would be reductions in wages for some workers and even unemployment. If such reductions were significant, social instability could be seriously undermined.

Not surprisingly, just as workers' tempers heated, the newspaper cooled the temperature in mid-1992 with a lead article on the front page, "Smashing the 'Three Irons' Cannot Be Over-Simplified." Although the writer endorsed the rationale of the smashes, he reminded readers of three points: (1) the smashes were only part, not all, of enterprise reform; (2) it was problematic to smash the old practices without reconstructing something in their place; and (3) thorough deliberations should precede action, and implementation should be sophisticated.[38] Although stories about the smashes continued to appear in the press, their number diminished significantly.

Another development that deserves attention is the publication of surveys of workers' responses to the smashes. This was significant in several ways. First, it showed that the future of workers under the smashes had aroused social concern. Second, the smashes began to give way to more systematic studies of personnel management. Third, the so-called surveys, although crude in method and analysis, could at least provide readers with an impression of workers' attitudes. All the surveys came to similar conclusions: Workers supported the reforms but worried about job security, especially as the unemployment safety net was not yet ready in China. The "resettlement" of the displaced was their primary concern.[39] Gradually, the smashes disappeared from the press and were replaced by more in-depth debates on reforms of the operating mechanisms.

Although the issue of resettling the "redundant" workforce was raised with the smashes, it has proven to be a long-term struggle that has continued to plague defense industries. The problem reflects the origin of China's defense factories and other parts of the state sector: They were created not only as production units but also as small communities that had to provide job and social security for all members.[40] This tradition was so deep-seated among the people that the factories could not abandon their workers at all. Factory bosses did not have many options in dealing with the problems of iron rice bowls: They could strengthen political work to woo the workers, grant permission for the early retirement of aged or handicapped workers, put redundant workers on long leave, or provide so-called retraining. In some cases, they had to pay the "released" workforce subsidies if not full wages.

Thus, we have reason to believe reports in the press that many released workers left the factories satisfied with their treatment.[41] Speaking at the tenth anniversary of CAPUMIT in mid-1995, Jin Zhude claimed that "more than a decade of conversion ha[s] not led to any closing down of industries and unemployment, nor did it increase the [economic] burden of the country; in rough estimates, a population of 12 million was stabilized."[42] This was an achievement in maintaining social stability, but it highlighted the dilemma for China's defense industrial base and for all state-owned enterprises posed by modernization and reform: stability or reform, caution or breaking new ground. In fact, the Ministry of Personnel promulgated a "Tentative Regulation on Dismissals in State-Owned Enterprises" only in early 1993.[43] After that, discussions on dismissals gradually diminished.

Perhaps the most important benefit of the smashes was the accelerated growth of the tertiary industry (*sanchan*) and the business activities of army factories, which had been used to accommodating redundant workers.[44] This distorted the development of this sector, however. A complaint in the newspaper highlighted three deficiencies: (1) Businesses lacked planning and coordination; (2) the structure was unbalanced—most of them rushed to open restaurants and convenience stores while other services were neglected; and (3) national assets drained away due to improper accounting.[45] Given the shortsightedness of factory bosses, these service industries became little more than vehicles without their own sound business prospects.

Enterprise Reforms

Deng Xiaoping's urge for bold steps in reform had brought not only rash decisions but also well-planned and down-to-earth reforms, most notably the stock share system and the individualized management system (*fenchang zhi*). Admittedly, the development of the stock share system has been slow, as it touches upon the sensitive issue of property rights. Although the first factory began experimentation in 1987, government authorities were still discussing the concept and reviewing the experiences in 1994. For example, Chengdu, in which many Third Line plants are situated, put its first defense factory under the stock share system only in 1993.[46]

The concept of individualized management appeared in *China Defense Conversion* as early as 1991 but was given only scant mention.[47] In contrast to the stock share system, however, it concerned only internal operating mechanisms and thus developed much more rapidly. It has basically two meanings: first, the coexistence—in one defense plant—of the conventional management system and the system used in enterprises with foreign funding, and second, separate management for production lines or branch factories producing civilian and military goods.[48] In practice, it took a multitude of forms—"one factory, two/multiple systems," dual production lines, and others.[49] The individualized management system was apparently popularized in 1992 when many factories were put under pilot reforms, and it quickly won approval among government authorities. The Hunan COSTIND urged all defense factories in the province to adopt it at a meeting in 1992, and Ding Henggao, in a special article to exhort managers to follow Deng's call for "changing the mindset" (*huan naojin*), encouraged defense industries to implement the new management system.[50] As reforms deepened, the individualized management system gradually gave way to the modern enterprise system (*xiandai qiye zhidu*).

The first salvo was fired by a deputy general manager of the China Ordnance Industry Corporation, Tian Ruizhang, who wrote as a special commentator for *China Defense Conversion* on 3 May 1994. In his opinion, defense industries under a modern enterprise system should be characterized by six points: (1) management of civilian and military production is separate; (2) property rights are clearly defined; (3) investors enjoy rights and liabilities in proportion to their capital contributions; (4) units entrusted with military production view national security as the primary interest and strictly abide by military contracts; (5) arms exports and civilian production aim at the promotion of productivity and economic efficiency; and (6) a scientific leading and organizational management system should be established to regulate the relationship among owners, managers, and workers.[51]

This commentary was followed by eleven more articles by contributors from authorities concerned with the defense industrial base, from COSTIND and the State Planning Commission to the China Aerospace Corporation.

The commentaries essentially centered on the themes in Tian's article. Interestingly, however, the closing article in the series called for caution. Written by Li Yintao of the National Defense Science and Technology Information Center, it argued for the unity of "military, social and economic efficiency." The cautious tone may be best summarized in his principle of "Experimentation with Resolution but Not After Quantities, Good Results Wanted but Not Chaos" (*jianjueshi, buqiuduo, zhiqiuhao, buyaoluan*), which was also the title of his essay.[52] Nevertheless, following a conservative approach characteristic of reforms since 1978, pilot schemes were initiated in some factories.[53]

Thus far, we have seen how China's defense industries have adapted to the economic setting in the process of conversion. Although the effectiveness of these reforms in transforming enterprises into economically efficient units remains in question, the adverse effect on the military has become clear.

The "Guarantee of Military Interests"

The reason for devoting a special section to the impact of defense conversion on the military is obvious: The protection of military interests is a cornerstone of China's defense conversion program. Moreover, as shown in coverage by *China Defense Industry,* the erosion of military interests has been a growing concern.

The Economic Calculus

Although it may be justified by Chinese communists as a form of dialectical process, the goal of establishing an economically sound defense industrial base conflicts with large-scale, high-speed military production, which signifies not only low profits but also almost always losses. Military production during peacetime is characterized by requirements for superior quality, specialization in technique, small quantities, production on short notice, and rapid changes in models and products. These imply an additional burden of maintaining technical teams and special machinery and storage of equipment not in use.[54]

The experiences of many factories demonstrate that they had to subsidize the research and development and production of military goods with revenues from civilian production. Moreover, they had to accept orders for military equipment even though they knew such orders would cause losses.[55] The Jialing Machinery Plant, famous for producing motorcycles, reported "serious losses" in military production from 1980 to 1990, except for 1981 and 1983 when there was "modest profit." In addition, the factory spent over 5 million yuan on the research and production of military goods.[56] Ironically, the Jialing experience was acclaimed as a model of protecting military interests.

Considering this heavy economic burden and the responsibility system—which pegs the salary of managers to factory performance—it is not difficult to understand why scarce capital and labor are transferred to civilian production. In some cases, plants have refused to accept orders for military goods. In exploiting the separate management of military and civilian production, some factories have simply shifted the burden of providing social welfare to all employees to military production branches.[57]

Brain Drain

In late 1992, when the smashes fever faded, a concern for the loss and aging of human resources in military production arose. The poor economic efficiency and the hostile working environment of most defense industries provided enough motivation for staff to enter other economic sectors; the decline in military production also led to substantial redundancies of technical staff. The situation was aggravated after the popularization of the separate management system. The issue of the poor salaries of technicians became increasingly prominent.[58]

The workers' union and even the political and military leaders were well aware of these facts. They knew research scientists and technicians could not be retained at defense factories and that new blood was not being attracted. Surely, the authorities suggested, the obvious solution to this problem was to improve the working environment; revamp recruitment, promotion, and vocational training; introduce material incentives; and, of course, intensify political work.[59] They could do little to implement these suggestions, however. In view of the lack of human resources, the quality of military products could not be guaranteed.

Quality in Jeopardy

In the early 1990s, there were calls for defense industries to improve the quality of their products; however, these calls concerned civilian goods faced with keen competition in domestic and international markets.[60] In mid-1995, there was a remarkable increase in the reports and discussions on upholding the quality of military goods. The profit-making calculus of the defense industries had led to difficulties in military procurement. A naval representative to the plants complained: "The adverse impact of the market economy on naval equipment development was no longer theoretical; the phenomena of 'guarantee profit, not military interests' (*baoli bubaojun*), 'guarantee quantity, not quality' (*baoliang bubaozhi*), and 'guarantee quality, not price' (*baozhi bubaojia*) happened in varying degrees among some factories."[61]

This problem became a hot topic at a conference on quality in national defense industries in Xi'an on 13 June 1995. Although it was not the first conference on this issue (the previous one was in 1993), it was given more coverage in the press. *China Defense Industry* not only reported the conference but also published all speeches by major participants, including the minister and vice ministers of COSTIND as well as deputy general managers of various defense industrial corporations and a vice minister of electronics. More important, this conference was held against a background of military accidents caused by quality problems in military equipment. Huai Guomo pointed out that "some factory bosses spent considerable energy on developing civilian production and promoting economic efficiency but became relaxed on quality control."[62]

The ordnance industry system responded by convening its own conference on quality assurance. Stories described how factories did good work in monitoring the quality of military production. In coordination with directives from the central government, a competition in quality control was also organized and review exercises were published in *China Defense Industry* for days.[63]

Conclusion

From the reports in *China Defense Industry* and the evolution of the newspaper itself, we have seen that China's defense industries in the process of conversion have been integrated gradually into the civilian economy despite tremendous difficulties. Deng Xiaoping's inspection trips in 1992 provided an impetus to defense conversion after it had been bogged down in the political and economic doldrums that followed the 1989 Tiananmen Square crackdown. The "smashes" fever failed to rebuild a replacement for what it intended to destroy but successfully raised a series of issues and put defense conversion back on a dynamic path. Again, enterprise reforms had to follow classic Dengist-style experimentation before full implementation; social stability was obviously placed above everything, even economic interests.

However, both defense industries (and their employees) and the military are victims of lost time. The human resources and quality problems cannot be solved without fundamental reforms. As long as the military goods pricing system is not overhauled, defense plants will not be interested in accepting military contracts. If the military must pay more for equipment, where will it obtain the money? Because of years of deficits it will not come from the central government, so the only possible solution would be extrabudgetary funding. This lack of money might erode the army's fighting capability, however. Obviously, defense planners have no easy choices.

The adverse impact on military production returns us to our discussions of the concept of defense conversion. The Chinese view conversion as a double-edged sword that will solve the problem of economic efficiency and promote military capability, yet a decade of defense conversion has resulted in a decrease in military production capacity (and, potentially, military capability). The development of dual-use technology in the future might be able to partially compensate for the loss. To date, however, the Chinese program has worked according to the conventional wisdom of economics: Resources essentially flow one way—toward profits, and from military to civilian markets.

Notes

The author wishes to thank the two editors, John Frankenstein and Jörn Brömmelhörster, as well as the staff of the Universities Service Centre at the Chinese University of Hong Kong, for their support and assistance. Appreciation also goes to Tai Ming Cheung, who generously lent his newspaper collection to the author.

1. *Zhongguo Jungong Bao* [China Defense Industry, hereafter *JGB*], 1 July 1994, p. 1; *JGB,* 4 July 1995, p. 1.

2. The term "defense industries" is used throughout to refer to industries and plants, primarily civilian-run, engaged in the production of military goods. Those owned by the army are specified as "army factories."

3. *Renmin Ribao* [People's Daily], 11 June 1985, p. 1.

4. *JGB,* 4 July 1995, p. 1.

5. *Zhongguo Jungong Bao* [China Military Civilian Production Integration, hereafter, *JZMB*], 1 January 1991, p. 1.

6. The Editorial Committee, *Zhongguo Baokan Dachuan* [Comprehensive Catalogue of Chinese Newspapers and Periodicals], vol. 1 (Beijing: Zhongguo Youdian Chubanshe, 1994), p. 46; *JGB,* 18 July 1995, p. 1.

7. *JGB,* 4 July 1995, p. 1.

8. Xie Guang, ed., *Dangdai Zhongguo De Guofang Keji Shiye* [Contemporary China's National Defense Science and Technology Undertakings], vol. 1 (Beijing: Dangdai Zhongguo Chubanshe, 1992), pp. 176–177.

9. *Renmin Ribao,* 16 August 1990, p. 6.

10. *JZMB,* 4 January 1994, p. 3.

11. *JGB,* 1 July 1994, p. 2.

12. *JZMB,* 1 January 1991, p. 1.

13. Please refer to the speech of Huai Guomo, a deputy director of COSTIND, at a *JZMB* conference about news reports, in *JZMB,* 8 September 1992, pp. 1 and 3. For the role change, see *JZMB,* 1 January 1993, p. 1.

14. *JZMB,* 1 January 1993, p. 1. For the series of essays on Deng's speeches, please refer to the newspaper from August to October, 1992.

15. *JZMB,* 4 January 1994, p. 2, and 29 July 1994, p. 3. Please refer to various issues of *JGB* in 1995 for the content of reforms in that year.

16. *Renmin Ribao,* 25 January 1994, p. 1; "The Politics and Economics of the Chinese Press," *China News Analysis,* no. 1505 (1 March 1994), pp. 1–3.

17. Lloyd J. Dumas, "Military Research and Development, and Economic Progress: Of Burdens and Opportunities," *Bulletin of Peace Proposals* 19:3–4 (1988), pp. 299–301; Kenneth L. Adelman and Norman R. Augustine, "Defense Conversion: Bulldozing the Management," *Foreign Affairs,* no. 2 (spring 1992), pp. 26–47.

18. "China's Ten Years of Conversion in Defence Industry—Speech Delivered by Mr. Huai Guomo at the Moscow International Conference on Conversion: Economic Readjustment in the Era of Arms Reduction," *Conmilit,* no. 165 (October 1990), pp. 19–22. See also his later speeches: *JZMB,* 22 October 1991, pp. 1 and 3; *JZMB,* 7 April 1992, p. 3; and other reports: *JZMB,* 13 August 1993, p. 3; *JGB,* 4 July 1995, p. 1.

19. *JZMB,* 1 March 1994, p. 1.

20. Ibid.

21. "Preface" to conference materials, 1993 Hong Kong Conference on International Cooperation to Promote Conversion from Military to Civilian Industry; Zhou Jianshe, *Guofang Ziyuan De Nixiang Kaifa—Zhongguo Junzhuanmin Wenti Yanjiu* [The Peaceful Development of Resources of National Defense—The Conversion of Military Resources to Civilian Fields in China] (Changsha: Hunan Chubanshe, 1992), pp. 52–55; *JZMB,* 8 January 1991, p. 1; *JZMB,* 16 July 1991, pp. 1 and 3; *JZMB,* 29 September 1992, p. 3; *JZMB,* 12 January 1993, p. 1; *JZMB,* 15 February 1994, p. 1.

22. A private source reports that a military regional (MR) commander confided that earnings from PLA activities were divided into thirds: one-third was kept by the unit, one-third went to the coffers of the military region, and one-third was sent to Beijing. The MR commander is reported to have said that the final one-third was sufficient to keep Beijing's nose out of the MR's business. See also Cheng Kuaile, "Zhunque Bawo Houqin Shichang Xingwei De Jiben Tezheng" [Correct Understanding of the Characteristics of Logistics Marketing Behavior], *Junshi Jingji Yanjiu* [Military Economics Studies], no. 12 (1994), pp. 53–54; *Zhongguo Qingnian* [Chinese Youth], no. 2 (1993); Ka Po Ng, "China's Defence Budgeting: Structure and Dynamics" in *China Review* 1995, eds. Lo Chi Kin et al. (Hong Kong: Chinese University Press, 1995), pp. 17–18.

23. Mel Gurtov, "Swords into Market Shares: China's Conversion of Military Industry to Civilian Production," *China Quarterly,* no. 134 (June 1993), pp. 213–215; Yitzhak Shichor, "Military to Civilian Conversion in China: From the 1980s to the 1990s," *Peace and the Sciences* (December 1993), pp. 36–37; Paul Humes Folta, *From Swords to Plowshares? Defense Industry Reform in the PRC* (Boulder, Colo.: Westview Press, 1992), pp. 2–6.

24. *JZMB,* 24 March 1992, p. 1; *JZMB,* 30 June 1992, p. 1; *JZMB,* 19 January 1993, p. 3; *JZMB,* 2 February 1993, p. 3; *JGB,* 16 June 1995, p. 1; *JGB,* 18 July 1995, p. 1.

25. Gurtov, "Swords into Market Shares," p. 214.

26. *JZMB,* 12 November 1991, p. 3; *JZMB,* 23 July 1993, p. 3; Ka Po Ng, *China's Defence Modernization Since 1977,* diss., University of Hong Kong (1994), pp. 139–141.

27. *JZMB,* 7 January 1992, p. 1; *JZMB,* 19 January 1993, p. 3; *JGB,* 2 June 1995, p. 1.

28. *JZMB,* 12 November 1991, p. 3; Liu Hongji, *Zhanzheng Dongyuan Xue* [War Mobilization Studies] (Beijing: Guofang Daxue Chubanshe, 1992), p. 252; You Qianzhi et al., *Zhongguo Guofang Jingji Yunxing Fenxi* [Analysis of China's Defense Economic Operations] (Beijing: Zhongguo Caizhen Jingji Chubanshe, 1991), p. 104.

29. Jacques S. Gansler, *Affording Defense* (Cambridge, Mass.: MIT Press, 1989), pp. 273–282; Julian Cooper, "Conversion Is Dead, Long Live Conversion!" *Journal of Peace Research* 32:2 (1995), pp. 129–132; Ng, *China's Defence Modernization,* p. 122; Fang Min and Fan Gongsong, *Da Guofang* [The Great National Defense] (Shanghai: Shanghai Renmin Chubanshe, 1994), pp. 228–234.

30. Sun Zhenhuan, *Zhongguo Guofang Jingji Jianshe* [China's Defense

Economic Development] (Beijing: Junshi Kexue Chubanshe, 1991), pp. 26–34; Folta, *From Swords to Plowshares?* pp. 81–99.

31. *JZMB,* 1 January 1991, p. 3.

32. For example, *JZMB,* 29 January 1991, p. 1; *JZMB,* 2 April 1991, p. 1; *JZMB,* 14 May 1991, p. 3.

33. *JZMB,* 1 January 1991, p. 1.

34. *JZMB,* 15 January 1991, p. 3; *JZMB,* 21 May 1991, p. 1; *JZMB,* 18 June 1991, p. 3; *JZMB,* 22 October 1991, p. 3.

35. *JZMB,* 15 January 1991, p. 3; *JZMB,* 22 January 1991, p. 1; *JZMB,* 5 February 1991, p. 3; *JZMB,* 14 May 1991, p. 3; *JZMB,* 31 December 1991, p. 1.

36. For the integration of military industries with local economies, see *JZMB,* 19 March 1991, p. 1 (Shanxi); *JZMB,* 2 April 1991, p. 1 (Sichuan); *JZMB,* 30 July 1991, p. 1 (Beijing); *JZMB,* 20 August 1991, p. 1 (Gansu).

37. *JZMB,* 7 April 1991, p. 1; *JZMB,* 21 April 1991, p. 1. For other reports on "Smashing the Three Irons," see also *JZMB,* 14 April 1992, p. 1; *JZMB,* 21 April 1992, p. 3; *JZMB,* 19 May 1992, p. 1.

38. *JZMB,* 16 June 1992, p. 1.

39. *JZMB,* 23 June 1992, p. 3; *JZMB,* 8 September 1992, p. 3. The newspaper carried one more similar survey on 9 February 1993, p. 3, but it was specifically about young workers.

40. Jiang Baoqi, ed., *Zhongguo Guofang Jingji Fazhan Zhanlüe Yanjiu* [A Study on China's National Defense Economic Development Strategy] (Beijing: Guofang Daxue Chubanshe, 1990), pp. 87–88; Peng Min, ed., *Dangdai Zhongguo De Jiben Jianshe* [Capital Construction in Contemporary China], vol. 1 (Beijing: Zhongguo Shehui Kexue Chubanshe, 1989), pp. 198–199; Ng, *China's Defence Modernization,* p. 131.

41. *JZMB,* 7 April 1992, p. 1; *JZMB,* 14 April 1992, p. 1; *JZMB,* 16 June 1992, p. 1; *JGB,* 20 June 1995, p. 1; *JGB,* 11 July 1995, p. 1.

42. *JGB,* 9 June 1995, p. 2.

43. *JZMB,* 5 February 1993, p. 3.

44. *JZMB,* 5 May 1992, p. 1; *JZMB,* 16 June 1992, p. 1; *JZMB,* 14 July 1992, p. 3; *JZMB,* 28 July 1992, p. 3; *JZMB,* 15 January 1993, p. 1.

45. *JZMB,* 22 January 1993, p. 3; *JZMB,* 5 February 1993, p. 3.

46. *JZMB,* 20 October 1992, p. 3; *JZMB,* 9 February 1993, p. 1; *JZMB,* 6 July 1993, p. 3; *JZMB,* 6 May 1994, p. 1; *JGB,* 19 July 1994, p. 3.

47. *JZMB,* 9 April 1991, p. 3.

48. *JZMB,* 28 April 1992, p. 3.

49. Examples of different forms of this system can be found in: *JZMB* 26 May 1992, p. 3; *JZMB,* 11 August 1992, p. 1; *JZMB,* 5 February 1993, p. 1; *JZMB,* 4 February 1994, p. 1.

50. *JZMB,* 28 April 1992, p. 1; *JZMB,* 28 July 1992, p. 3.

51. *JZMB,* 3 May 1994, p. 1.

52. The sequence of the 1994 commentaries in *JZMB* (after Tian Ruizhang's) was May 6, 10, 13, 17, 20, 24, 27, 31, and June 3, 7, 10.

53. *JGB,* 6 June 1995, p. 1; *JGB,* 11 July 1995, p. 3.

54. Adelman and Augustine, "Defense Conversion," pp. 27–28; You et al., *Zhongguo Guofang Jingji Yunxing Fenxi,* pp. 97 and 99; *JZMB,* 12 November 1991, p. 3.

55. *JZMB,* 2 June 1992, p. 1; *JGB,* 11 July 1995, p. 1.

56. *JZMB,* 14 May 1991, p. 1; *JZMB,* 24 June 1994, p. 1.

57. *JZMB,* 30 June 1992, p. 1; *JGB,* 12 August 1994, p. 1.

58. *JZMB,* 18 August 1992, p. 1; *JZMB,* 22 June 1993, p. 3; *JZMB,* 2 June 1994, p. 1; *JGB,* 18 July 1995, p. 1.

59. *JZMB,* 25 May 1993, p. 3; *JGB,* 2 June 1995, p. 1; *JGB,* 18 July 1995, p. 1.

60. *JZMB,* 29 January 1991, p. 3; *JZMB,* 21 December 1993, p. 3.

61. *JGB,* 23 June 1995, p. 1.

62. *JGB,* 16 June 1995, p. 1; *JGB,* 27 June 1995, pp. 1–2; *JGB,* 4 July 1995, pp. 1–2.

63. *JGB,* 7 July 1995, pp. 1–2; *JGB,* 11 July 1995, p. 2; *JGB,* 14 July 1995, p. 2; *JGB,* 18 July 1995, pp. 1–2.

PART 3

Case Studies

6

Converting the Military-Aviation Industry to Civilian Use

Yitzhak Shichor

Before and immediately after the communist takeover in 1949, China barely had the capability to develop and produce aircraft, although elementary aeronautical engineering research was carried out in a few universities with crude experimental facilities. The more modern plants, built by the Japanese in the 1930s in Manchuria, had been dismantled and removed by the Soviet Union in the summer of 1945. While this looting precluded the immediate resumption of postwar production, the buildings, the airfields, and the skilled employees remained and provided an infrastructure for the future development of China's aircraft industry.[1] This infrastructure also included a few small-scale and ill-equipped aircraft repair factories, abandoned by the retreating Guomindang. Some of these were immediately turned over by the Chinese Communist Party (CCP) to civilian use—an early example of military-to-civilian conversion—so that by October 1949, when the People's Republic of China (PRC) was proclaimed, only six aircraft repair factories had survived.[2]

Half a century later, China's aviation industry has become a complex system covering research and development, design, production, testing, training, and trade enterprises. By the mid-1990s it included about 140 aircraft and aircraft-related production enterprises, 37 scientific research centers on aviation, and a variety of higher education institutions and professional and vocational schools with a total workforce of 800,000, including some 190,000 engineers, technicians, and managers. Since the early 1950s this complex has produced more than 14,000 aircraft (mostly for military use), more than 50,000 aircraft engines, and more than 5,000 types of airborne equipment.[3]

Catering primarily to military needs, much of this potential was underutilized under Mao Zedong. After Mao, China's aviation industry began to face a real crisis. By the mid-1990s, following a period of promoting aircraft exports, Beijing had realized that its endlessly recycled and upgraded aircraft—based on outdated Soviet technology of the 1950s—could no longer satisfy the growing demand, in quantity and quality, of the People's Liberation Army Air Force (PLAAF), let alone that of foreign customers. The combination of reduced orders and the priority given to economic growth and modernization has led China's military aviation production to a forced landing.

In the early 1980s, China began to convert underutilized or useless military aviation production, research, and technological capacity to civilian aviation as well as nonaviation use in order to accomplish a new takeoff. Motivated by reduced threat perceptions, deradicalized politics, and highlighted economic growth, much of this effort has been carried out in cooperation with foreign countries. As a specialized military industry that concentrates on a relatively small number of products with a rather high visibility, China's aviation industry provides a good example—supported by adequate data—of Chinese-style military-to-civilian conversion. The first part of this chapter covers the organization, infrastructure, problems, and main characteristics of China's aviation industry under Mao. The second part deals with China's aviation industry after Mao, the conversion of its military potential to civilian use, and the implications of conversion for the modernization of the PLAAF.

China's Aviation Industry Under Mao

Built with massive Soviet support in the context of the Cold War in general and the Korean War and Indochina war in particular, China's aviation industry was shaped by Mao's revolutionary legacy and ideological militancy; by political upheavals, economic difficulties, and technological backwardness; and, above all, by Beijing's high threat perceptions and constant security concerns. It is no surprise, therefore, that aircraft factories and research institutes under Mao promoted primarily military production while paying rather limited attention to civilian needs.

Organization

Responsibility for aircraft production shifted considerably after the early 1950s. An Aviation Industry Preparatory Group, set up as early as 1950 within the Ministry of Heavy Industry, was reorganized in April 1951 as an Aviation Industry Bureau (AIB). In August 1952, aviation (and ordnance) industries were transferred to a newly created Second Ministry of Machine Building (MMB), sometimes referred to as the Ministry of Defense Industry. A Commission of Aviation Industry (CAI), established in April 1956, was responsible for research, design, and production of aircraft and missiles.

In February 1958 the Second MMB merged once again with the First MMB, now covering both military and civilian production. Later that year relations with Moscow began to deteriorate, making the Chinese even more determined to develop their own weapons. A major step in this direction was the establishment in October of the Commission for Science and Technology for National Defense (COSTND), which incorporated the CAI. On the basis of its suggestion to centralize all military research and development (R&D) and to reestablish quality control in military production—greatly eroded by the radicalism of the Great Leap Forward—the new Third MMB was created in September 1960 and assumed responsibility for all defense industries from the First MMB. All research and design related to aircraft were combined three months later under an Aviation Research Academy (also known as China's Aeronautical Establishment), renamed in January 1961 as the Sixth Research Academy.

To further increase efficiency and specialization, in 1963 military industries were split among three new MMBs (the Fourth, Fifth, and Sixth), so that the old Third MMB was now responsible for only the aviation industry, including missiles. In November 1964, a Seventh MMB (for missiles) was established. From then on, supervision of the Sixth Research Academy (aviation) shifted from COSTND and the People's Liberation Army (PLA) to the Third MMB, which now dealt exclusively with aircraft. This organizational structure remained virtually unchanged until the end of the Maoist era.

Infrastructure

China's aviation industry was officially founded on 17 April 1951. According to the "Resolutions on the Construction of Aviation Industry," eighteen factories (sixteen from the air force and two from the Bureau of Weaponry Industry) were transferred to the AIB, together with ten thousand employees. By the end of the year, six major aircraft factories had been established or expanded: two in Harbin (Heilongjiang), two in Shenyang (Liaoning), one in Nanchang (Jiangxi), and one in Zhuzhou (Hunan). About 80 percent of the personnel and 70 percent of the equipment were concentrated in these factories. By the end of 1952, China's aviation industrial base already covered thirteen factories with thirty thousand employees.[4]

Most of these factories were expanded and rebuilt with massive Soviet assistance, on which China's aviation industry depended much more than did other military production enterprises. The degree of this dependence was evident during the Sino-Soviet alliance and even more so after its dissolution. As early as July 1949, Liu Shaoqi asked for Stalin's assistance in modernizing China's armed forces and defense industry, primarily the air force.[5] Soviet equipment began to arrive in Shenyang in May 1951, followed by a first batch of twenty Soviet experts sent to assist in the reconstruction and expansion of China's six major aeronautical repair factories. An agreement to this effect was signed between the two governments in October 1951.[6]

Indeed, during the recovery period (1950–1952), six aircraft and aircraft engine repair factories were reconstructed and expanded.

Recently declassified Soviet documents record a conversation held in Moscow on 20 August 1952 on this issue. Responding to a request by Premier Zhou Enlai for Soviet aircraft, Stalin urged the Chinese to build their own aircraft manufacturing plants. Reluctantly, Zhou said that it would be very difficult to build aircraft, particularly jet aircraft, in the course of the First Five-Year Plan and noted that it would take at least three years to begin construction of engines and five years to begin construction of aircraft. Yet Stalin insisted. Pointing to the examples of Poland, Czechoslovakia, and Hungary, he offered to send engines and other aircraft parts so that China could set up assembly plants and, three years later, aircraft plants. Zhou remained lukewarm. Although he admitted that Beijing was beginning to understand this need and to organize maintenance and assembly plants, he added that "if comrade Stalin finds it necessary to hasten the process, then they will take all appropriate measures to comply" and would act upon the advice of the Soviet government.[7]

Indeed, in May 1953, shortly after Stalin's death, Moscow and Beijing signed a more comprehensive agreement that provided for Soviet assistance in 141 major projects (increased in 1955 to 156), including forty-one major national defense industrial enterprises–of which thirteen, or almost one-third, were aircraft factories.[8] China imported from the Soviet Union the manufacturing technology of seven types of aircraft, nine types of airplane engines, five types of tactical missiles, and several hundred types of airborne equipment. In addition to 847 Soviet aviation experts who worked in China, hundreds of Chinese workers and students were sent to the Soviet Union and Eastern Europe. China took advantage of this Soviet assistance, combining theory with practice, learning with production, and schools with factories.

Aided by Moscow, nineteen new aviation-related factories were built during China's First Five-Year Plan (1953–1957). All of them were located in major cities with convenient transportation facilities; accessibility to human, material, and energy resources; and good geological and hydrological conditions. New factories were built alongside the old ones, which allowed China to construct and produce simultaneously and to use the method of "one director for two factories."[9] This method was later used in military-to-civilian conversion.

In the meantime, China's aviation industry began expanding through civilian-to-military conversion. As early as April 1951, the CCP gave priority to the construction of aviation industries, which were provided with all available manpower and resources. From 1952 to 1954, thousands of workers, technicians, leading cadres, and executives were transferred to the aviation industry from party, administration, railway, communications, arms industry, car assembly factories, and PLA units. By the end of 1957, China's aviation industry had 100,000 employees. Up to 1955, a number of steel, rubber, paint, chemical, and textile factories were also transferred from the various ministries to the aviation industry. Two factories were taken over from the Civil Aviation Bureau.[10] Thus, by 1959–almost ten years after its establishment–China's aviation industry had made great progress. Its research and design institutes and twenty-nine major manufacturing enterprises were capable of producing piston-engine trainers, transport aircraft, and even jet fighters.

In fact, it was in the late 1950s that China's aviation industry began to develop the overcapacity that could be rather easily converted to civilian use in the 1980s. Investment in the aviation industry during the Second Five-Year Plan (1958–1962), which converged with the Great Leap Forward, increased three times compared to that in the First Five-Year Plan, and the number of employees jumped to 300,000. Under the slogan of "building two factories with the investment of one and at half the time," plans were made to double the number of military factories in three years, and in 1960, construction on more than twenty factories began at the same time. Of the fifty-one projects initiated between 1958 and 1960, none reached production during the Second Five-Year Plan and twenty newly started projects had to be halted. This large-scale construction led to an enormous waste of resources, to serious quality problems, and to disorderly management.

During the three years of readjustment (1963–1965), steps were undertaken to correct this situation and by the early 1960s, the expansion and reconstruction of most aircraft-related factories had been slowed or stopped. Nevertheless, two sets of aircraft and engine factories in Chengdu and Xi'an and two airborne-equipment factories in Lanzhou, for which construction had begun in 1960, were completed and commenced production by the end of 1963. At the same time, to fill the blanks in China's aviation production system caused by the Great Leap Forward, a number of old aviation factories in Beijing, Changchun, Harbin, and Nanjing were reconstructed and expanded. In addition, several local civilian factories were appropriated to ensure the development and production of aircraft and missiles.

As mentioned above, most of the aircraft and aircraft-related factories were located along the coast in the traditional industrial centers. Although some inland construction of aviation industries began in the 1950s and early 1960s, only in October 1964 did these experiments warrant a policy incorporated into the "1965 National Defense Industry Work Plan." Accordingly, in early 1965 the Third MMB decided to stop or curtail construction of First and Second Line aviation enterprises and shift some of them to the inland Third Line (*sanxian*). Within two or three years, forty thousand workers and staff had to move to outlying mountain areas. Much of this effort was concentrated in Guizhou, where a complete set of aviation factories was built. About two thousand workers and staff from aviation factories in Shanghai, Tianjin, and Beijing were ordered to Guizhou, together with all the equipment that was to be installed for production within six months. Similar bases were built in the 1970s in Shaanxi, Jiangxi (a set of helicopter factories), Hubei (a water-based aircraft factory), and Hunan. At the same time, more than ten factories in Shanghai were reconstructed and expanded for developing and producing large civilian airliners, an initial step toward conversion.

This buildup gathered momentum during the Cultural Revolution, when many factories were constructed without planning, feasibility studies, or official approval. From 1966 to 1976, more than forty aviation industry projects were established. This large-scale duplication further increased overcapacity, and new factories competed with old ones for "food." Moreover, most of these Third Line factories were located in poor and remote mountainous areas and lacked adequate transportation and runways. Many had to be closed later, as production capacity could not be utilized and enormous problems could not be overcome. Huge amounts of funds were wasted: Investment in Third Line aviation industry reached 93.4 percent of the total investment in the aviation industry in the Third Five-Year Plan (1963–1967) and 83.2 percent in the Fourth Five-Year Plan (1968–1972).[11]

In sum, by the late 1970s, post-Mao China had inherited an oversized yet underdeveloped military aviation industrial infrastructure. The attempts by China's moderate leaders, such as Zhou Enlai, Ye Jianying, and, primarily, Deng Xiaoping, to reorganize, rehabilitate, and consolidate China's aviation industry were frustrated by the radicals.

Military Production

The development of China's aviation industry has undergone three stages: repair, copy-production, and independent design. As early as 1952, Zhou Enlai reportedly urged China's embryonic aviation industry to develop the capacity to copy-produce aircraft in three to five years. This was a quick, practical, and feasible way of supplying the armed forces with reliable weapons for urgent needs, before independent design and original models could be developed. In this way, advanced technology was digested and absorbed, product design and manufacturing techniques and skills were quickly mastered, people were trained, and management was improved. This policy of imitation and the early dependence on the Soviet Union have affected China's aviation industry to this day.

In July 1954, the Nanchang Aircraft Factory (established in May 1951) turned out the PRC's first copy-produced piston-engine YAK-18, the CJ-5 (*chujiao,* or primary trainer). A few days later, Mao Zedong congratulated the workers and staff on their "triumphant success" and added: "This is an excellent beginning for the founding of our nation's airplane building industry."[12] From 1954 to 1958, a total of 379 CJ-5 trainers were produced. By that time, the development of the CJ-6, begun in Shenyang, was also transferred to Nanchang and production began in 1962. By 1986, 1,796 CJ-6s were produced. In the meantime, the Shenyang Aircraft Company concentrated on fighters and, in July 1956, the final assembly (under a licensing agreement) of China's first MiG-17F subsonic fighter, known as J-5 (*jianji-ji,* or fighter plane), was completed. A total of 767 J-5s were produced from 1956 to 1959, followed by modified versions produced since the 1960s by the Chengdu Aircraft Factory. Some are still in service with the PLAAF. Beijing urged research institutes and factories that had acquired basic aircraft manufacturing skills through copy-production to start the design and production of an indigenous Chinese aircraft. This was the JJ-1 (*jianjiao,* or fighter-trainer), China's first jet trainer turned out by the Shenyang Aircraft Factory in July 1958. Yet China's attempts to produce high-altitude, high-speed fighters (later known as J-6 and J-7, based on the MiG-19 and MiG-21, respectively) failed. Although Chinese-made MiG-19s and J-6s were certified for production in 1959, as a result of the Great Leap Forward a large number of aircraft produced in 1959 could not be delivered because of inferior quality and had to be improved for three years. The J-6 and J-7 production lines were left unfinished and inoperative. Nonetheless, Beijing decided to proceed.[13]

The J-6, whose production had been resumed in early 1961, made its first flight by the end of the year.[14] Within two years the Shenyang Aircraft Factory finally managed to produce "good quality" J-6 supersonic interceptors. Several modifications were later introduced and, by Mao's death, some 270 Chinese J-6s had been exported, mostly to Pakistan, North Vietnam, and North Korea. Production of the J-7, begun at Shenyang in 1964, proved to be more difficult. In 1961 (after the Sino-Soviet break), China signed a licensing agreement with the Soviet Union for the production of the MiG-21 and received technical data, complete aircraft and engine kits, some equipment, and raw materials not available in China. Some data was not provided, however, causing difficulties in production. China claimed that production "progressed very fast." The J-7 made its maiden flight in January 1966 and was certified for production in June 1967.[15] Since 1964 and 1965, production of the J-7 has been simultaneously carried out by three aircraft factories: Shenyang, Chengdu, and the newly established Guizhou Aircraft Production Base. The Chinese may have succeeded in overcoming some J-7 production problems after they had allegedly managed to derail an unknown number of late-model Soviet MiG-21s, shipped to Vietnam through Chinese territory.[16]

Production of the H-6 (*hong,* for bomber), a licensed Chinese copy of the Soviet TU-16 medium jet bomber, began in the late 1950s, using a complete set of technical data, kits, and two samples according to Sino-Soviet agreements reached in 1956–1957. Assembled by the Harbin Aircraft Factory, the H-6 made its initial flight in September 1959. From 1961 to 1964, all H-6 production shifted to the Xi'an Aircraft Factory, where, in December 1968, the first Chinese-made H-6 began its test flights. By

September 1964, the initial model was retrofitted in Xi'an into a nuclear bomb carrier and was used for the first time in 14 May 1965 to deliver a 20–40 kiloton (KT) atomic device in China's second nuclear test. On 17 June 1967, it delivered a 3 megaton (MT) thermonuclear bomb (China's first test of its kind), and on 17 November 1976, it delivered a 4 MT thermonuclear device, China's largest test to date. By then, nearly half of China's nuclear tests (ten out of twenty-one) had been carried out by the modified H-6.[17] Batch production of the H-5, based on the Soviet light jet bomber IL-28, began at the Harbin Aircraft Factory in 1967 after three years of trial production using incomplete design drawings. A modified H-5 was used on 27 December 1968 to deliver a 3 MT thermonuclear device.

While relying heavily on Soviet models in aircraft production, the Chinese also managed to develop genuine prototypes. One example is the Q-5 (*qianjiji,* or ground attack aircraft), a radical modification of the MiG-19 undertaken by the Nanchang Aircraft Factory since 1958. Designed to match state-of-the-art foreign ground attack aircraft, the Q-5's development was suspended in 1961 due to accidents, the Great Leap Forward, and China's consequent economic decline. Although the Q-5 was flight-tested in June 1965, batch production began only in late 1969 after four years of problem solving and considerable improvement. A modified nuclear bomb carrier, the Q-5A, made its first flight in August 1970. On 7 January 1972, it delivered a 20 KT hydrogen bomb in China's thirteenth nuclear test.

Although based on the experience of copy-producing the J-7 and a study and analysis of "documents of foreign fighters obtained," the J-8 (another Chinese-designed aircraft developed at Shenyang) was less successful.[18] A high-altitude and high-speed fighter, its concept and specifications were worked out in late 1964 after years of preparation and deliberation. Trial production of the J-8, however, was seriously disrupted by the Cultural Revolution, especially during its violent phase in July-August 1967. Despite all difficulties, the J-8 made its first test flight in July 1969; nonetheless, it took more than ten additional years to overcome many technical and management problems before the J-8 was finally certified.

Although helicopter production technology was introduced in China in the mid-1950s, by the end of the 1970s China's aviation industry had managed to produce only one type: the Z-5 (*zhisheng feiji,* or helicopter), based on the Soviet MI-4. Although it was certified in December 1959, mass production of the Z-5 began only in 1963 after the end of the Great Leap Forward. By 1979, 545 Z-5s had been produced. The development of two other helicopters was terminated—the Z-6 primarily because of technological difficulties, and the Z-7 mainly because of economic difficulties.[19]

In sum, considering the internal and the external circumstances, by the time of Mao's death in September 1976 China's aviation industry and research establishment had achieved great progress in copy-production, modification, and even independent development of various military aircraft. Yet these achievements had been undermined by serious shortcomings, related to excessive dependence on the Soviets, backward technology, difficulties in adapting theories and designs to actual production, economic scarcity, increased isolation from the international community, and, above all, radical politics.

First, the Great Leap Forward in the late 1950s forced the unprepared industry to quickly produce new kinds of aircraft. The most dramatic upheaval came in the 1960s and 1970s; according to the Chinese themselves, the Cultural Revolution "severely interfered [with] and destroyed the smooth development of Chinese military aircraft." In ten years, more than thirty projects were launched and canceled as a result of mismanagement. "Large amounts of development investment paid for nothing; valuable time was wasted, resulting in heavy losses." Technical problems led to many accidents, and hundreds of aircraft had to be returned to the factory for repair.[20] These constraints confused and complicated R&D plans, created unnecessary competition, stretched technical and production capacities to their limits, caused delays in other development and production programs, seriously compromised quality, and led to a drain on state resources.

Civilian Production

Although China's aviation industry concentrated on military production for twenty-five years, by the late 1950s some progress had already been made in the production of transport aircraft that, although also used by the PLAAF, were mostly used for civilian purposes. They included copies and modifications of foreign models—such as the Soviet IL-14, YAK-12, YAK-16, AN-2, AN-12, AN-14, and MI-4, or the Czech Super-Aero 45—designed and produced in Harbin, Nanchang, Xi'an, Shaanxi, Shanghai, Shijiazhuang, and Shenyang, especially since 1957–1958.[21]

A noteworthy example is the Fengshou utility biplane, virtually a direct Chinese reproduction of the Soviet AN-2, later known as Y-5 (*yun,* for transport). From 1958 to 1986, 948 Y-5s were designed and produced under a licensing agreement in Nanchang (727) and, since 1970, in Shijiazhuang (221), including 229 agricultural and 114 passenger versions. The Y-5 has been one of the Chinese aviation industry's most popular models. It has been extensively used for civilian cargo and passenger transportation, as well as for agricultural and geological purposes. Similarly, among the 545 mostly PLAAF Z-5 helicopters produced in China by 1979, 86 were passenger versions, 7 were agricultural and forestry versions, 13 were rescue versions, and 2 were aerial survey versions.[22] Indeed, as shown in Table 6.1, by 1960 civilian aircraft accounted for 17 percent of China's total aircraft production.

Table 6.1 China's Aircraft Production, 1950–1960

	Military	Percent	Civilian	Percent	Total
Fighters	767	100	–	–	767
Trainers	278	94	17	6	295
Transports	41	17	199	83	240
Total	1,086	83	216	17	1,302

Sources: Adapted from *Dangdai Zhongguo De Hangkong Gongye* [China Today: Aviation Industry] (Beijing: Zhongguo Shehui Jiaoxue Chubanshe, 1988), p. 40; Charles Barton, "China's Growing Airlines and Aviation Industry," in *China's Economic Dilemmas in the 1990s: The Problems of Reforms, Modernization, and Interdependence,* ed. Joint Economic Committee, Congress of the United States (Armonk, N.Y.: Sharpe, 1992), pp. 473–474.

In fact, as early as 1956, when the production capability of China's aviation industry began to increase, the issue of combining military and civilian production began to be discussed. Mao mentioned it in January and again in April, calling for learning two sets of skills: "The skill of producing civil products in military industry, and the skill of producing military products in civil industry." A civilian production management organization was established by the Second MMB and, in early 1958, a meeting on civilian production held by the AIB adopted the policy of "production of both military and civil products simultaneously on the basis of ensuring the fulfillment of military production and gradually raising the level of technology."[23] Following this policy, China's aviation industry began to produce metal cutting machine tools, generators, steam turbines, automobiles, tractors, and motorbikes. Thus, the Zhuzhou Aero-Engine Factory has been producing motorcycles and motorcycle engines since the mid-1950s and for many years was China's major supplier of heavy motorcycles for both military and civilian use. Consequently, the share of civilian products in the total output of the industry increased considerably and in 1960 reached more than 40 percent.[24]

After the early 1960s, however, civilian production declined considerably. According to the Chinese, this was a direct outcome of the Great Leap Forward: "Incorrect criticism to civil production affected the implementation of the principle of combining military production with civil production for a long time." Due to "the errors of leftism," aviation industries were criticized as "unprofessional" for producing civilian goods and were told "to mind their own business" (i.e., to produce military aircraft). No one was ready to take responsibility for civilian production and, consequently, by 1978 its share in the output value of China's aviation industry fell to 6.5 percent. More than half of the industry's production capacity had remained unutilized while the rest concentrated on military production.[25]

Nevertheless, the domestic political radicalization was only one reason for this change. Other reasons must have been the exacerbation of the Sino-Soviet conflict and the simultaneous, growing U.S. intervention in Southeast Asia. Involved, directly or indirectly, in a series of regional conflicts and feeling isolated and besieged, Beijing maintained a climate of war fed by high threat perceptions until the early 1980s. Cut off from all external sources of military supply, the Chinese had to promote the production of military aircraft at the expense of civilian ones in constant preparation for an imminent war.

It was only toward the end of the 1960s that China resumed its development of civilian aircraft. A noteworthy example is the Y-7, based on the Soviet AN-24. Begun in October 1966 at the Xi'an Aircraft Factory, the Y-7 made its first flight in December 1970. It then underwent rigorous tests for more than a decade before being certified. Also developed in Xi'an, the Y-8—a heavier payload, medium-sized transport based on the Soviet AN-12—made its first flight in December 1974. In 1972 trial production was switched to the Shaanxi Aircraft Factory, a Third Line base established in the mid-1960s. The Y-8 was certified for serial production in February 1980. A small number of Y-9 medium-range transports were also produced after the early 1970s and are still being used for military and civilian purposes. Development of the Y-10 airliner and the Y-11, a small transport, began in the mid-1970s. None of these transports were produced in large quantities in Mao's time, when the conversion of China's aviation industry to civilian use had stalled.

This lag resulted from three kinds of interrelated problems, some unique to China's aviation industry but most of them common to all Chinese industries, both military and civilian. The first concerns the domestic political situation and its dialectical relationship with the international environment. In retrospect, it is clear that Beijing tends to attribute the failures of its aviation industry primarily to those revolutionary upheavals that swept China following the late 1950s. The Great Leap Forward mentality of overanxiety, wanting quick results at the expense of quality, led to mass production even before aircraft were certified and to "heavy setbacks and losses. A large number of finished aircraft could not be delivered and the aircraft built in one year required three years to repair."[26]

> Seeking more and fast, exaggerating [the] subjective role without counting [the] practical possibility[ies] prevailed in [the] aviation industry. Arrogance and overanxiety with quick results were developed . . . stressing the method of "trial production at a fast pace." A lot of serious quality problems happened due to careless large scale process changes, blindly reducing the use of necessary tools and fixtures and standard parts and the cancellation of many approved procedures and inspection operations. There was not a single qualified aircraft or engine of the J-6 and Z-5 aircraft delivered from 1958 to 1960.[27]

Then came the Cultural Revolution. Again, large numbers of aircraft, engines, and missiles could not be delivered due to quality problems: "It was calculated that during the 'great cultural revolution' development of more than 30 types of aircraft was stopped, [and] large amount of development investment paid for nothing; as for the cost of time, it could never be compensated."[28] As mentioned above, the blind buildup of Third Line aviation factories actually reduced rather than increased aircraft production potential, leading to an enormous waste of materials, personnel, and funds.

China's failure to build sound and solid military aircraft thwarted the experiments in civilian production. The circumstances of growing threat perceptions and international isolation, especially after the late 1950s and early 1960s, further fed domestic political radicalism and created a sense of insecurity and instability that could by no means support or legitimize conversion. Divided on many fundamental issues, China's leadership could not unite around a firm strategy of converting excess defense production capacity because the "civilian" government failed at times to control the military and also to a certain extent merged with it, both personally and ideologically.

Other reasons concerned the economic system. Maoist China's centralized socialist command economy provided few or no incentives for innovative and creative economic thinking, including the diversification of production through a more rational use of resources. As mentioned previously, attempts at civilian production by defense industries were considered less prestigious than military production and invited leftist criticism. In addition to Mao's principles of self-reliance, China's exclusion from the international economic system ruled out the possibility of direct foreign investment and considerably limited export opportunities. Because arms were provided to other countries free of charge, there was also no extra income to be spent on civilian (or military) production.

Finally, civilian conversion could not take off because of the technological backwardness of China's military industries, despite later Chinese claims that one of the primary incentives for conversion was the rather advanced military-industrial technology.[29] The Sino-Soviet conflict inflicted considerable damage on China's air power due to its dependence on Moscow, something the Chinese are reluctant to admit even today. In August 1960, Moscow canceled all contracts and withdrew its experts, who stripped their offices and removed all blueprints and designs. In the early 1960s, the PLAAF operated at a low rate because of the lack of spare parts; work on more than ten types of new aircraft as well as most aircraft factories slowed or halted. By the mid-1960s, the number of combat-ready aircraft dropped drastically, and many aircraft had to be cannibalized in order to provide spare parts and ammunition to maintain the others.

Isolated from the rest of the world, China had no substitute for Soviet technologies on which its military and, primarily, aviation industries depended. This traditional emulation of external models stifled creativity and ingenuity and underscored the difficulties of adapting scientific theories to practice. Mao's inspired revolutionary upheavals exacerbated these problems. On the one hand, scientists and engineers preferred to keep a low profile for fear of criticism, demotion, or outright purge. On the other hand, the authorities forced them to undertake premature projects that consequently failed. Chinese sources tend to ascribe these failures "to unrealistic specifications which were beyond China's available technology and industrial capability at that time."[30]

Maoist China's technological backwardness was also an outcome of the economic system, which relocated the enterprises' depreciation funds and profits to the center instead of using them (or a portion) for technological transformation and renovation. Overall priority was always given to capital construction, leading to a waste of resources on superfluous inland buildup even in times of economic shortages. The deployment of China's aviation industry all over the country further undermined its technological level. About 150,000 young staff and workers who joined the aviation industry during the Cultural Revolution had not received proper technical training and were not competent.[31] Consequently, the development of a new generation of aircraft was hindered, leaving the PLAAF with outdated equipment.

By 1976, after sixteen years of production, the J-6 could still not be replaced by the J-7, in itself based on obsolete technology. As the Maoist era expired, "the gap between China's aviation industry and the advanced aviation industry in the world had become greater."[32] Under these circumstances of technological backwardness, domestic political instability, external threats, and economic centralism, military-to-civilian conversion was indeed inconceivable. Nevertheless, most of these circumstances have begun to change since Mao died, paving the way for a far-reaching transformation in many fields, including military-to-civilian conversion.

China's Aviation Industry After Mao

The share of military production in China's aviation industry output value, which by the end of Mao's era had accounted for about 95 percent, gradually declined after the early 1980s. Correspondingly, the share of civilian production has increased from about 6 percent in 1978 to 13 percent in 1980, 40 percent in 1985, 61 percent in 1986, 64 percent in 1990, and about 75 percent by the mid-1990s. The value of China's aviation industries' civilian output during the Seventh Five-Year Plan (1986–1990) was 3.76 times higher than that of the Sixth Five-Year Plan (1981–1985).[33] Nonetheless, such expansion and diversification have hardly affected the buildup, let alone the design and R&D, of China's military air power.

Adaptation

While emphasizing military production as a first priority, in 1978 Mao's successors, led by Deng Xiaoping, began to realize that defense industries had failed to fully use their production potential, which could be converted to satisfy civilian needs. Under prior conditions of constant alert for war, defense industries had been overstructured, overequipped, and overstaffed. With the relaxation of international tension, the decrease by a large magnitude in state orders, and the emphasis on modernization and development, this surplus production capacity could be redeployed to serve national economic construction. Indeed, Deng Xiaoping—then chairman of the CCP Central Military Commission (CMC)—pointed out in early 1982 that all departments in the defense industrial sector, apart from fulfilling the requirements of research and production of arms and military equipment, should spare no efforts to research, develop, and produce civilian products.[34] Following this strategy, China's aviation industry began to transform the purely military production structure into a combined military and civilian production structure and gradually switched to civilian production guided by the laws of the market economy.

An early organizational step in this direction, undertaken by 1979, was the establishment of the China Aero-Technology Import-Export Corporation (CATIC) as the principal, though not exclusive, trading arm of China's aviation industry. Enterprises were allowed to select their own export channel. At that time the Chinese government also decided to start selling aircraft, after several thousand aircraft had been provided free of charge to seventeen countries from the 1950s to the 1970s. Contracts were now signed between CATIC and foreign countries and between CATIC and the enterprises. Earnings in 1980 were 2.46 times more than in 1979, and the value of exports in 1986 was 5.5 times more than in 1979. The exporting enterprises were able to retain some of the foreign exchange; the total amount, the size of profits, and the performance rate of the contracts in turn determined their bonuses. Moreover, as an incentive they could keep a higher share of foreign exchange earned from the export of civilian goods.

Selling aircraft abroad was merely one aspect of the change in the aviation industry budgeting system. For a long time, new aircraft development funds had been allocated to the relevant administrative organizations by the government, which would reimburse them for their spending, overlooking cost-effectiveness. As in other countries, this policy led to uncontrolled expenditure and waste of scarce resources. After 1981, Beijing introduced the budget responsibility system and the staged bonus method, in which the rights and liabilities of each party in the research, design, and production process are made clear. Each part of the contract had to be reviewed and approved to reach the optimal results in terms of marketability and profitability. This called for further organizational changes.

In May 1982, Beijing abolished one of the military production MMBs and reformed and renamed the others, thus painting them with a more "civilian" color. The Third MMB became the Ministry of Aeronautics Industry. Another step to encourage conversion was taken in 1984, when enterprises were given greater decisionmaking power in planning, finance, labor, distribution, supply, sales, and foreign trade. Whereas military aviation production was still centrally governed, civilian aviation and nonaviation production would be adjusted by the market. In 1985, the capital for developing civilian aircraft was included in the state plan for the first time. In addition, capital owned by the enterprises, as well as bank loans and government subsidies, could be partly used for civilian production. These new financial resources have enabled China's dated aviation industry to undertake for the first time the technological transformation necessary for modern civilian production and cooperation with foreign companies.

Domestic cooperation with other ministries and departments, as well as within the ministry, has also increased considerably while departmentalization, so characteristic of Mao's China, was gradually abandoned. Beijing claims that by 1986 aviation industry enterprises had established 273 combined enterprises with 698 organizations and agencies all over China. As early as 1979, for example, the separate Shenyang aircraft production, design, and research enterprises joined to establish the Shenyang Aircraft Company, while the separate Shenyang aero-engine production, design, and research enterprises joined to establish the Shenyang Aero-Engine Company, which in turn set up six subsidiary factories (for gas turbines, aluminum profiles, gas meters, etc.).[35]

This tentative process of reform received a boost after 1986, when the civilian aspects of the Ministry of Aeronautics (along with those of the other three former military production ministries) were placed directly under the State Council, while the military aspects (including R&D, production, science and technology, and trade) were placed under the administration of the Commission for Science, Technology, and Industry for National Defense (COSTIND), created in 1982. Thus, despite the alleged decline in military aviation production, these organizational changes, the stable political environment, improvements in management, greater sensitivity to the market, and attempts at quality production led to an increase in total output value, which in 1986 was 73.8 percent higher than in 1978. The period (1979–1986) that covered the Sixth Five-Year Plan (1981–1985) was one of the most fruitful in the history of China's aviation industry. Also in those years, Beijing established about eighty new aviation industry colleges and universities, postgraduate institutes, secondary technical schools, and workers' classes, not only "to ease the short term need for qualified personnel" but also as "a back force for the future development of aviation industry."[36]

Meeting with aviation industry experts in early 1986, Premier Zhao Ziyang pointed out that the concept of military-civilian combination was completely different from the simple concept of civilian production by military enterprises.[37] Indeed, in 1987 Beijing decided to further reform the defense industry system and integrate it completely into the national economy. "Development policy changed from [the] previous combination of military and civil production with military production as the main task to the combination of military and civil [production] with civil production as main task." It was also reaffirmed that although aviation production should be done according to a central plan, civilian production would be open and intersectoral.[38]

To ease this integration, in April 1988 the Ministry of Aeronautics Industry merged with the Ministry of Astronautics Industry to form a new Ministry of Aeronautics and Astronautics (MAA). To further "release" the forces of production and following similar changes in other sectors, however, in June 1993 the combined ministry was split into two corporations: the China General Aviation Industry Corporation (CGAIC) and the China Aerospace Industry Corporation (CAIC). A state-owned enterprise, CGAIC consists of more than two hundred scientific research, production, and business operations units with a total workforce of over 560,000. It is responsible for all aspects of Chinese aviation, for military as well as civilian purposes, including the formulation of development and business strategies, the supervision of operation and management, the supply of technologies and equipment, quality control, test flights, and foreign affairs.

Shortly afterward, in August 1993, CATIC became the core company of a newly established CATIC Group, which incorporated thirty-three other related companies. By that time, CATIC had twelve offices all over China, thirty-nine offices in twenty-six countries overseas, and trade relations (including the export of some 650 aircraft of different types) with over seventy countries and regions valued at more than U.S.$7 billion. In early 1995, its international representation increased to forty-five agencies in thirty countries. According to Liu Guomin, chairman of the board of directors of CATIC Group, its establishment was aimed at helping to concentrate human, material, and financial resources and promoting the export of China's high-tech products in the international market.[39] In addition to the import and export of aviation and nonaviation products and technologies, both military and civilian, CATIC is engaged in production, design, and development and provides a variety of engineering, labor, publishing, and other services. By 1994, at least four aviation industrial groups had been established, each composed of corporations, companies, and factories—some considered closed-tier enterprises and some semi-closed-tier enterprises. The Shanghai and Xi'an aviation industrial groups include about a dozen enterprises each—mostly at home but also abroad—and the Guizhou Aviation Industrial Group has some forty-eight different enterprises.

In sum, as mentioned above, in Mao's China aviation industries and research organizations always produced civilian aircraft (and other goods) in addition to fulfilling their quotas for weapons and military equipment. From 1978, however, the proportion of civilian production to military production has reportedly increased enormously. Typical of Chinese communism, practice preceded theory. After long preparation, it was as late as 1988 that the various defense industry ministries submitted their respective production capability adjustment plans: Of the total original production capacity, roughly one-third was said to be enough to satisfy the revised military needs, while two-thirds could be switched to support national economic construction.[40]

Conversion

This overcapacity—both human and material—has now been diverted to civilian production primarily, but by no means only, in the field of aviation. Former "military" aviation enterprises have now been classified into two kinds of military-civilian combination enterprises—those that are mainly, but not only, engaged in aviation products and those that are mainly, but not only, engaged in nonaviation products—and into two kinds of enterprises that are totally engaged in either aviation or nonaviation products. This policy led to immediate results. From 1978 to 1986, the output value of China's aviation industry civilian products reached an average yearly growth rate of 37.4 percent, and the number of products increased from more than two hundred to more than three thousand.[41] These included civilian aircraft, parts, and equipment, which are necessary for the conversion of China's aviation industry.

Although Mao's China had developed a number of civilian aircraft, few were produced on a large scale. Now, Beijing launched new projects and continued those already begun in Mao's time. Some ventures, such as the abortive U.S.$300 million attempt to reverse-engineer a copy of the Boeing 707 (known as the Y-10), failed. After two Y-10 prototypes had been built and flight-tested in September 1980, "the development program was terminated because of reasons of market and cost."[42] Other attempts have been more successful. Since the early 1980s, many passenger planes have reportedly been delivered by China's aviation industry to domestic airline services. For example, by late 1994, seventy-three Y-7 passenger planes delivered to customers since January 1984 were flying 148 domestic air routes, thus becoming their principal workhorse. In the 1980s, the Y-8, Y-11, Y-12I, and Z-8 were also batch-produced and put into service.

Other civilian aircraft included the Mifeng (Honeybee) and Qingting (Dragonfly) ultralight models; the Q-2 and AD-100 ultralight aircraft produced with two U.S. companies; the Mifeng-6 airship, used mainly for erecting high-voltage transmission lines; the WZ-5 high-altitude reconnaissance pilotless aircraft; and the D-4 miniature remotely piloted vehicle (RPV). Although these aircraft have also served military purposes, they have been used mostly for aero-seeding afforestation, forest-fire prevention, crop seeding, fertilizer spreading, weeding, insect elimination, mice killing, disaster relief, aero-photography, wild-animal tracking, offshore oil-rig support, geological prospecting, short-distance shuttle passenger and cargo transportation, archeological surveys, and so on.

Since the early 1980s, China's aviation industry has also taken on civilian projects not directly related to aviation—by providing other ministries and industries with spare parts, special equipment, instruments, and services. For example, one-third of the work completed in 1985 by the Ministry of Aeronautics Industry's Fourth Establishment dealing with design and construction had little or nothing to do with aviation. In addition, China's aviation industries supplied industrial gas turbine power-generating sets derived from aero-engines to the oil industry; cooperated with the Ministry of Coal Industry to develop a coal mine safety monitor system; built a television set assembly line for the Tianjin Television Factory; designed bicycles, foodstuffs, and textile factories; and, in cooperation with the Textile Industry Ministry, developed and produced six types of spinning and weaving equipment. Reportedly, all these achievements have reached advanced world standards, considerably raised production efficiency, and increased revenues.

Yet most conversion has been achieved by China's aviation industry itself, which, during the Sixth Five-Year Plan (1981–1985), established about sixty civilian production lines. These included textile and light industry machines, medical equipment, automobiles, motorbikes, bicycles, washing machines, and refrigerators. Some brand names, such as Jincheng and Nanfang motorcycles, Changhe and Songhuajiang minibuses, Shenfei tourist buses, Kaidi bicycles, and Chankong filling machines, have become well known.

China accomplished much of this development alone. Yet, after many years of isolation from the international community, Beijing finally abandoned self-reliance. Commencing sales of aircraft in 1979, China's aviation establishment has also become involved in nonaviation projects abroad. In 1980, for example, CATIC established a Design and Construction Engineering Company that soon became active, primarily in the Middle East (Jordan, Kuwait, and the United Arab Emirates). Most of the income has likely been reinvested in the aviation industry.

More importantly, to upgrade and diversify its aviation products and to introduce more advanced technology into the industry so as to enhance competitiveness, China's aviation industry began to import foreign technologies, research and test equipment, and promote cooperation with international aircraft producers such as Boeing, McDonnell Douglas (MD), Messerschmitt-Boelkow-Blohm (MBB), and Airbus. These imports of aviation technology and subcontracting agreements have opened new markets for China's aviation industry through compensation trade, generally valued at 30 percent of the deal. In 1979, China's aviation industry began to produce aircraft parts for thirteen foreign firms with which it has since signed some one hundred industrial production contracts.[43] By 1994, the total delivery value of these parts had reportedly reached over U.S.$200 million.

Some of these transactions involved the development and production of commercial jetliners. This plan is based on the assumption that "although China's civil aviation sector continues to grow at an annual rate of over 30 percent, it still fails to meet the demands of economic development and international and domestic tourism."[44] With more than forty airline companies and 420 planes (92 of which were acquired in 1994), China's civil aviation is expected to continue growing at over 20 percent a year so that by the year 2000, more than four hundred new passenger planes will be needed. In 1994, over sixty airports were under construction or expansion with a total of 120 (98 in 1992) in operation.[45]

The following examples provide a sense of the extent of China's conversion efforts from domestic and international perspectives.[46] China's most important aircraft production cooperation project began modestly in 1979 when the Shanghai Aircraft Factory (later to become Shanghai Aviation Industrial Corporation, SAIC) began producing landing gear doors for the MD DC-9. Following a decade of exploration and negotiation, in March 1985 the two parties signed a twelve-year licensing agreement for the production of twenty-five MD-82s with options on fifteen additional aircraft.[47] The fact that SAIC had less experience in military production than other Chinese aircraft factories is probably one of the reasons it was chosen by MD. Although more problems existed, the absence of deeply rooted traditional production patterns made it easier to solve them. Indeed, by 1994, thirty-five MD-82/83 jetliners had already been jointly assembled, five of which were sold back to the United States and thirty to China Eastern Airlines and China Northern Airlines. In that year, SAIC signed another contract for the joint production of forty MD-series jetliners (such as MD-82 and MD-83).

SAIC and CATIC have also negotiated with Boeing on the establishment of a joint venture to produce the rear sections of Boeing 757 and 737-700 jetliners. In 1988 the Chengdu Aircraft Corporation and MD signed a contract for the production of one hundred noses for the MD-80. Other civilian projects include the Y-8, a heavy-duty transport based on the Soviet AN-12 that has been produced by the Shaanxi Aircraft Factory since 1984–1985, incorporates imported electronic equipment from the United States, and has been improved since 1987 in cooperation with Lockheed; EC-120 helicopters; MPC-75 feeder planes (with MBB); and the Y-7H-500 passenger aircraft.[48] A new contract for the joint production of twenty medium-range MD-90-30 passenger planes worth U.S.$1.6 billion, to be completed by the year 2000, was signed in December 1994 between CATIC and MD. The project will be undertaken by four aircraft manufacturing plants, in Shanghai, Chengdu, Xi'an, and Shenyang.

First established in 1937, the Shenyang Aircraft Corporation (SAC) has turned out thousands of military aircraft since the mid-1950s. These include the J-5, J-6, J-7, and J-8 fighter, trainer, and reconnaissance aircraft, as well as the J-8II, one of China's most advanced high-altitude, high-speed, all-weather fighters. With the decline of military orders, however, SAC began to develop civilian products. These include several types of air-conditioned buses (introduced from Fuji Heavy Industries) and a state-of-the-art two-engine hovercraft with advanced control systems.

SAC has also established ties with some of the world's leading aircraft firms, such as Boeing, Saab, and MBB, and has signed contracts to manufacture tail cones for C-130 Lockheed transport aircraft, rudders for British Aerospace, a variety of engine parts for General Electric, and Dash-8 cargo doors for Canada's De Havilland (which later became part of Boeing). SAC invited foreign companies to jointly carry out research, development, and production of airplane parts, automobiles, hovercraft, die forges, and machinery. For example, it plans to invest 100 million yuan (U.S.$17.2 million) in a light-duty cross-country jeep project with a designed annual production capacity of ten thousand jeeps and an expected annual value of 800 million yuan (U.S.$137.9 million).[49]

Another example of military-to-civilian conversion comes from Guizhou province, where many Third Line industries were built in the 1960s and 1970s. In 1985, the China Jiangnan Aviation Industrial Corporation (CJAIC) was the first to use imported parts to assemble Toyota lightweight trucks. In the following three years, CJAIC developed and manufactured its own vehicles of various models (under the trademark of Hangtian), including minivans, ambulances, police cars, instrument cars, equipment carriers, armored vehicles, and control tower command cars. Some 80 percent of the parts used in Hangtian vehicles are produced domestically. To improve its accessibility to the transportation networks and to the markets, CJAIC decided to relocate its light automobile production base to Maocaopu, a provincial-level high-tech development zone in Zunyi City.

Since the early 1970s, the Guizhou Aviation Industrial Group Corporation (GAIGC) has designed and produced the J-6A supersonic all-weather fighter; the JJ-7 supersonic fighter-trainer (also known as FT-7); and, together with the Chengdu Aircraft Company, the J-7III, an advanced high–medium altitude, high-speed, all-weather, day-and-night fighter. After 1986, however, GAIGC began to use its high-tech facilities to develop over eight hundred kinds of auto parts, whose output value in 1992 soared to 300 million yuan. The truck sealing compound produced by GAIGC's Hongyang Machinery Factory filled a domestic void and is now used by over 150 motor vehicle manufacturers in China; it has a 60 percent share of the market.

Employing over seventy thousand people in forty-eight enterprises and possessing thirty thousand pieces of equipment—some of them imported—GAIGC also produces meat-processing machinery, cigarette-packing machine lines, marble-granite mining and processing equipment, cutting and measuring tools, combined jigs and fixtures, nonstick cookware, and various pumps and valves. To improve its accessibility to the transportation networks and the markets, GAIGC built a processing production base for auto parts and basic machinery parts in Xiaohe, a provincial-level, high-tech development zone in Guiyang City.[50]

With sixty-five hundred employees, the Changhe Aircraft Industries Corporation (CAIC) in the eastern suburb of Jingdezhen, Jiangxi province, is China's leading center for research and production of large helicopters. As early as 1975 it began to develop civilian products. To adapt itself to the requirements of a market economy, it also introduced advanced technology and equipment from abroad to set up four production and assembly lines for civilian goods, primarily cars, spare parts, and components. So far, about twenty types of cars have been designed and produced, including a variety of buses and minivans. In 1994, more than sixteen thousand small cars (and two Z-8 helicopters) were turned out.

Developed since 1975 on the basis of the French Aerospatiale SA.321JA Super Frelon, the Z-8 was test-flown for the first time in 1985 and was finally certified November 1994. A large, multipurpose transport helicopter with a maximum takeoff weight of 13 tons, it is the heaviest helicopter designed and manufactured in China itself. It can be used for cargo and personnel transportation, rescue and medical aid, forest fire control, and ocean search and rescue patrols. "To ensure the continuance of the Z-8 programme the Jiangxi Helicopter Factory adjusted its products in time and carried out a policy of *'supporting the military development programmes with the profit earned in civil production'* so that the income from the civil product sales could be used to supplement the limited Z-8 programme budget" (emphasis added).[51] As a result of these achievements, CAIC output value in 1994 increased by 32 percent and sales income by 23 percent, compared to 1993. CAIC planned to invest 250 million yuan so that by 1995, production capacity of minivans would reach seventy thousand.[52]

Reportedly, the conversion of the Harbin Aircraft Manufacturing Corporation (HAMC), a long-time producer of China's H-6 bombers, began in early 1980 when very few orders had been received—all of them for spare parts. For the first time in many years, not even one aircraft was ordered. One solution was international cooperation. In 1980, CATIC signed a licensing agreement with Aerospatiale and Turbomeca for production of the French AS365/NI Dauphin II helicopter, to be known as the Z-9. Produced by HAMC in cooperation with the Zhuzhou Aero-Engine Company, the Harbin Engine Factory, and the Baoding Propeller Factory, the Z-9 made its first flight in 1982 and was certified in June 1987. The fifty Z-9s produced by 1995 have been used for various civilian purposes as well as for public security. Since 1982, HAMC has subcontracted for the production and assembly of a variety of parts for the British Aerospace BAe-146 and SD3 series.

Another solution to the reduction of orders was self-reliance. Raising its own funds and taking its own risks, HAMC started to develop the Y-12 using Pratt and Whitney engines. On 14 July 1982, the Y-12 made its first flight and was finally certified in December 1985. In 1985, the Y-12's interior and air-conditioning system were retrofitted by the Hong Kong Aircraft Engineering Company (HAECO). Production began in 1986. After two years of tests by the British Civilian Aviation Authority (CAA), on 20 June 1990 the Y-12 was awarded an international airworthiness certificate. Of China's civilian aircraft, it is the most successful export item thus far. By March 1993, when it was on display at the Paris International Air Show, fifty Y-12s had been sold, and by late 1994, seventy-two Y-12s had been delivered to domestic as well as foreign customers in seventeen countries and regions. Its certification by the U.S. Federal Aviation Administration in March 1995 reportedly led to a flood of orders, one hundred by a Japanese businessman and two hundred by two U.S. companies.[53] Among other things, the Y-12 has been used for mineral detection.

HAECO, a partner to the Hua Nan Aerotechnology Corporation (a joint venture with CATIC), is also cooperating with the Xi'an Aircraft Company (XAC) to produce the Y-7-100, a retrofitted and modernized version of the Y-7 that was certified in 1986. More advanced Y-7-200s, incorporating a number of Western components, are also being developed by XAC.[54] As early as September 1980, it signed a contract for the subassembly of Canadair CL-215 aircraft, followed since early 1982 by a number of contracts with Boeing for the production and subassembly of a variety of parts for Boeing 737 and 747 airliners. To undertake these contracts, many production lines had to be reengineered. XAC has also begun building the Eaglet-100, a high-quality, low-price, small aircraft that can fly at a maximum speed of 110 kilometers per hour for 135 minutes with two people (or a takeoff load of 338.5 kilograms) using no more than twenty meters of simple runway. Priced at about U.S.$13,000, it can be used for sports, flight training, aerial photography, and tourism.[55]

Takeoff

Since the early 1990s, the total output value of China's aviation industry has grown considerably. From 8.6 billion yuan in 1991, it grew to 11.02 billion yuan in 1992 (a 28.0 percent increase over 1991), of which 7.86 billion yuan (71.0 percent) consisted of civilian production. In 1993 it reached 14.96 billion yuan (a 35.7 percent increase over 1992), of which 11.00 billion yuan, or 73.53 percent, represented the value of civilian production (see Table

6.2). Sales income hit 15.3 billion yuan, of which 11.20 billion yuan (a 43.0 percent increase), or 73.2 percent of the total income, came from civilian goods. By early 1993, eight thousand kinds of civilian goods were turned out by two hundred civilian production lines in China's aviation industries. Some of the aviation industry enterprises made large profits, and many of those that once registered losses began to produce profits.

Table 6.2 China's Aviation Industry Output, 1979–1993 (in billions of yuan and percent)

Year	1979	1980	1981	1982	1983	1984	1985	1986	1987	1988	1989	1990	1991	1992	1993
Total	2.80	3.96	3.19	3.80	4.75	5.80	4.54	3.64	4.52	5.745	–	–	8.60	11.02	14.96
Military	2.33	3.44	2.49	2.89	3.69	4.28	2.72	1.46	–	1.723	–	–	–	–	–
Percent	83.2	86.9	78.1	76.1	77.7	73.8	59.9	40.0	–	30.0	–	–	–	–	–
Civilian	0.47	0.52	0.70	0.91	1.06	1.52	1.82	2.18	–	4.022	–	–	–	7.86	11.0
Percent	16.8	13.1	21.9	23.9	22.3	26.2	40.1	60.0	–	70.0	–	64.0	–	71.3	73.5

Sources: Adapted from Paul Humes Folta, *From Swords to Plowshares? Defense Industry Reform in the PRC* (Boulder: Westview Press, 1992), pp. 230–231, and official Chinese data.

Similarly, aviation industry exports have increased considerably. CATIC's 1993 export volume of over U.S.$400 million accounted for nearly half of the U.S.$870 million total exports of the entire industry. It included civilian aircraft and mechanical and electronic products valued at U.S.$550 (other sources mention $539 or $560) million, or 63.2 percent (compared to U.S.$500 million in 1992 and U.S.$20 million in 1985), nearly one-third of China's total exports of machinery and electronic products in that year. In the first half of 1994, aviation industry machinery and electronics export value reached U.S.$257 million. These exports are expected to increase by U.S.$200 million a year during China's Ninth Five-Year Plan (1995–2000) to finally reach U.S.$1.8 billion.[56]

Since 1994, the recently established CGAIC has begun to implement an ambitious ten- to fifteen-year "takeoff plan" aimed at creating a huge aviation industrial group that will adopt more flexible business mechanisms, develop and expand civilian production, adjust to the international market to increase imports and exports, and reach a higher technological level with its aviation and nonaviation products. With expected annual growth rates of 18 to 25 percent, income from sales in the year 2000 should reach 60 billion yuan—nearly four times the 1993 figure—and output value should reach 50 billion yuan, compared to nearly 15 billion yuan in 1993.[57]

The strategic plan covers specific developmental goals in ten areas, one of which (to be discussed later) is military aircraft. Other areas include civilian aircraft (R&D of primary aircraft as well as development and improvement of the Y-7, Y-8, and Y-12 series through international cooperation); advanced helicopters (apparently for anti-tank, night warfare); gas combustion engines (primarily for military-use turbojet engines); airborne equipment (with emphasis on the importation of aviation electronics systems); experiments, test flights, and long-term research (for new types of military aircraft and large-scale civilian airliners); tertiary industry; and imports and exports (trade volume will rise to U.S.$3 billion by the year 2000). For example, according to a three-step civilian aircraft development plan, MD planes would be produced in cooperation with the United States; a 100-seat plane would then be developed through international cooperation, and, finally, a 180-seat plane would gradually be produced. It is expected that during China's Ninth Five-Year Plan a solid base for civilian aircraft production at international standards will be built.[58]

Finally, the takeoff plan anticipates a great expansion of domestic and foreign markets as well as increased attraction of domestic and foreign capital for high-quality nonaviation products. These include minibuses, medium-size and luxury buses, motorcycles, air-conditioner compressors and air conditioners, compressors for refrigerators, timepieces, liquid crystal displays, and computer-controlled equipment. The plan predicts that the industry will earn at least U.S.$800 million in 1995 by exporting aviation and nonaviation products and that exports will account for a third of the industry's total output value by the year 2000. CGAIC will also participate in the development and construction of the Shanghai Pudong New Area, which, by the year 2000, will become one of the major export-processing industrial bases for China's aviation industry.

Although China's aviation industry could rely on its so-called high-tech advantages in its military-to-civilian conversion efforts, it has always suffered (as has the entire defense-production complex) from a chronic shortage of funds. To overcome this disadvantage, the newly established CGAIC—with the support of several overseas companies and banks—set up the China Aviation Foundation. As soon as it was launched, the foundation entrusted the Hong Kong–based Tienlee Company with raising foreign capital for sixty-six carefully selected industrial projects, including technical upgrading, purchase of equipment, personnel training, marketing studies, or even for use as circulating capital.

Reportedly, this initiative has proven to be very successful. Whereas before the foundation was established China's entire aviation industry had received no more than U.S.$7 million of overseas investment, within less than a year the new foundation had managed to raise U.S.$96.6 million in special funds and foreign direct investment. By 1995, U.S.$200 million had reportedly been raised. Using these funds, eight new enterprises have begun to operate on new bases, among them Tianjin Tianli Aviation Electronics Ltd., which has received U.S.$7.95 million for new product development.[59]

CGAIC president Zhang Hongbiao has said that the corporation was hoping to receive annual Japanese government loans of U.S.$100 million in the second half of the 1990s to buy Japanese aviation technology and equipment. These loans were meant to compensate for China's aviation industry for spending more than U.S.$700 million on imports of Japanese software, equipment, and key parts since the mid-1980s, while Japan bought virtually no Chinese aviation products. He added that China was hoping to receive similar credit from other governments that sell a good deal of aviation equipment and products to China (such as the United States, Germany, Britain, France, Italy, and Spain), despite the Organization for Economic Cooperation and Development's (OECD's) restrictions on governmental loans. To raise more funds CGAIC, or one of its major subsidiary companies, may be transformed into a shareholding company (in which the state still controls the majority share) that would be listed on international exchange markets and issue stocks overseas.[60]

Military Implications

Military-to-civilian conversion, the priority accorded to overall economic development, the drastically reduced threat perceptions, and the alleged reduction in arms orders (to be discussed later) have by no means affected the determination of China's post-Mao leaders to modernize its defense system in general and its military aviation in particular. Some of the first steps undertaken by Mao's successors were to restore order, and primarily quality, in China's military R&D and production and to establish many new defense technology and industry institutes.

Significantly, after twenty years of isolation and for the first time in its history, the PRC has the option of acquiring weapons and military technologies from the West and, since the early 1990s, also from Russia. Soon, however, the Chinese realized the limits of this option. After a spell of arms window-shopping in the late 1970s, Beijing came to realize that the enormous costs, the technological incompatibility, the risk of renewed dependence on external suppliers, and its own sense of relative security precluded a wholesale import of aircraft and other weapons. In July 1980, Nie Rongzhen wrote to Liu Huaqing, then deputy chief of the general staff in charge of weapons and military equipment, saying that it was impossible to import modernization and that China should develop its own weapons and military equipment while importing only a few foreign technologies.

Monitoring aircraft imports has been one of COSTIND's main responsibilities. A few months after its creation, it organized a group of specialists to undertake a feasibility study on the import of military aircraft. Its report, submitted to the CMC in December 1982, concluded that "on the existing basis of domestic aeronautical science, technology and industry, China should import from foreign countries certain critical technologies and actively modify the aircraft in service and develop new types of aircraft."[61]

Thus, despite China's open-door policy in other fields, "self-reliance" remains as the cornerstone of its military development. Released from the bonds of Maoist radicalism, China's aviation industry has intensified its efforts to design and produce advanced military aircraft.

In 1977, the CMC had already decided to accelerate the development of a ground attack aircraft, a medium high-altitude fighter, and a medium short-range bomber and the modification of existing aircraft. The Q-5I and Q-5IA, improved and extended-range versions of China's Q-5, were certified in December 1983 and January 1985, respectively. Following "a firm order" received in April 1981, an export version—the Q-5III—was designed. From January 1983, "several dozens" of Q-5IIIs were exported (by the late 1980s, at least two hundred were sold, twenty to Myanmar, forty to North Korea, and 140 to Pakistan). Moreover, in an early example of the use of Western technology for military purposes, a contract signed in July 1986 between the Nanchang Aircraft Company and Aeritalia resulted in a retrofitted version with improved maintainability, reliability, navigation accuracy, penetration capability, and combat effectiveness.[62] Designated Q-5D, it made its first successful flight on 30 August 1988.

Modification of China's medium-sized high-altitude fighter also began in 1977. The J-7II, a further improved version of the J-7, was certified in September 1979 not only for use by the PLAAF but also for export. Of the different export versions, the J-7M has been the most successful. Incorporating some fifteen improvements of domestic as well as foreign origin (avionics, airborne radar, and a heads-up display from GEC, Marconi, and Smiths), it was certified in November 1984 and has been exported since 1985. The JJ-7, a Mach-2 fighter-trainer version also derived from the J-7II, was flight-tested in July 1985 after six years of design and trial production by the Guizhou Aircraft Company. It was certified in February 1988. Finally, the J-7III, an all-weather day and night fighter jointly designed and produced by the Chengdu and Guizhou Aircraft Corporations, was completed in 1981. Trial production started in 1982. Flight-tested in April 1984, it was certified in February 1988.

In March 1976 the State Council and the CMC decided to accelerate the development of the J-8, begun in 1964. A daytime fighter that could not operate in bad weather, its design was officially certified only in March 1980. By that time it was already outdated: Since 1976, a J-8I all-weather operation fighter had also been in the development stage. After a number of failures and improvements it was finally certified in July 1985. Although the J-8I was considerably more advanced than the J-8, it still suffered from serious defects. Consequently, an improved version, designated as J-8II, was completed by the Shenyang Aircraft Corporation in March 1984 and was finally certified in October 1988.

In October 1976, the CMC approved a new development program for retrofitting the H-6 with a second-generation navigation and bombing system. The design was certified in August 1977, and the retrofitted H-6 made its first flight in March 1981. Its most recent modification is the H-6D, a first-generation air-to-ship missile-carrying bomber. Undertaken by the Xi'an Aircraft Factory together with the Ministry of Ordnance and the Ministry of Electronics, the H-6D has two YJ-6 (*yingji,* for eagle-strike) air-to-ship missiles, a missile-aiming fire control system, an automatic navigation and bombing system, a missile-heating system, and a new type of radar-detecting system. The H-6D was certified in December 1985 and the YJ-6 in 1986.

By that time China had also completed the flight-test program of the SH-5 seaplane and, in December 1984, it was officially delivered to the PLA navy. In 1987, the SH-5 was modified to meet requirements for forest fire prevention and firefighting, ocean exploration, and other agricultural and forestation purposes. This was a rare example of the direct conversion of a Chinese military aircraft to civilian uses.

Although it emphasizes the overall development of the civilian aspects of China's aviation industry, the takeoff strategic plan that has begun to be implemented by CGAIC by no means overlooks the military. Although it represents just one out of ten areas of development, military development also overlaps with some of the other fields and is quite considerable. For example, according to the plan, priority will be given to the research on and production of advanced fighters that reach the highest international standards, which may include the Super-7 (S-7) fighter, an updated version of the J-7 (but still based on the outdated MiG technology) that is scheduled to make its maiden flight in 1996. Research for its development has reportedly been conducted with Russia, Pakistan, and some Western countries. According to Western diplomatic sources, China had already received a number of confiscated Iraqi MiG-29 fighters from Iran in 1992 in return for missile technology and a nuclear power station. Developed by the Shenyang Aircraft Corporation, the Super-7 underwent test flights in mid-October 1994. Russian technicians have reportedly cooperated with the Shenyang Liming Engine Corporation in the development of its engine, an improved version of the MiG-29's RD-33, which was designed to provide the new Chinese jet fighter with 5 percent more thrust than the MiG-29. It is assumed that China will need another partner in order to introduce Western state-of-the-art technology essential for the development of the Super-7, after its cooperation with Grumman was suspended following the Tiananmen Square massacre.

In fact, China is in the process of improving another aircraft: the J-10, a multirole fighter that is lighter than the MiG-29. It has been developed since 1992, reportedly with Israeli assistance, by the Chengdu Aircraft Corporation and incorporates technology employed in the Israeli fighter

Lavi. Different in its appearance from the Lavi, the J-10 prototype is reportedly scheduled to take its first test flight in early 1996. Both new fighters are expected to become operational around the year 2000.[63]

Some—and probably many—of the funds needed for R&D and the production of these and other military aircraft came from arms sales. Although the Chinese often refer to the reduction of military orders as one of the main reasons for launching their military-to-civilian conversion policy, these claims hardly reflect the facts. From the early 1950s to the early 1990s, China reportedly exported 2,074 aircraft of different military types. Of these, 997 (48 percent) were transferred in Mao's time, spread over twenty-six years from 1950 to 1976. In the fourteen years after his death, China sold 1,077 aircraft (52 percent), more than half of them since the mid-1980s. In 1991–1992, when its military-to-civilian conversion effort was already in full swing, China's military aviation establishment received orders for an additional 173 fighters, including forty J-6s and twenty-one J-7Ms to Bangladesh, forty J-7s to Pakistan, and seventy-two J-7Ms to Iran.[64] Thus, so far there has been no real slowdown in China's aircraft exports.

For example, over 70 percent of Pakistan's inventory of some five hundred Chinese military aircraft have been bought since 1978 and 53 percent since 1986. Sino-Pakistani military cooperation continues and, in September 1994, Pakistan received the first batch of six K-8 (Karakorum) jet trainers. Worth more than U.S.$20 million, the trainers were designed and produced at the Nanchang Aircraft Manufacturing Company under a 1986 agreement (whose origins go back to 1983) between it, CATIC, and the Kamra Pakistan Aeronautical Complex. By that time, twenty-one K-8s had been produced.[65] Thus, by the late 1980s and early 1990s, while traditional customers such as Iraq could no longer buy Chinese military aircraft, old markets for such aircraft had been expanded (e.g., North Korea) and new markets had been opened (e.g., Myanmar). Military-to-civilian conversion could hardly have been motivated by "sharply reduced" external orders.

Conclusion

By the mid-1990s, conditions for the military-to-civilian conversion of China's aviation industry had never been better. Relative political stability provided a framework for a firm and consistent policy, an essential requirement for successful conversion. Reform of the economic system added another important element. The gradual transformation of China's socialist command economy into a socialist market economy has increased its ability to diversify production, generate funds, maximize profits, and participate in the international economy. All these have paved the way for technological renovation, improvement of management, and the more effective use of personnel as well as hardware.

Indeed, according to Chinese and some Western claims, China's military-to-civilian conversion has been a huge success—so much so that it is becoming a model to be emulated by other countries that are facing similar problems. China's aviation industry is a case in point. Within the last fifteen years or so, three-quarters of its overall production capacity, formerly used for military purposes, have apparently been converted to civilian ones. Nevertheless, there is little information about the economic viability of this process, particularly in view of the difficulties of state-owned industries in general (funding, losses, debts, redundancies, marketing, sales, accounting, pricing, profitability, inventories, etc.). Beijing has provided little information about these problems.[66]

A more fundamental observation is that most of this civilian production is an outcome of brisk economic expansion rather than real conversion. Unrelated to military production lines, new civilian production lines have been constructed or acquired from abroad. Like other state-owned industries, China's military industries have had a surplus of unutilized or underutilized workers, space, and machinery that could easily be converted to civilian use. Unlike civilian state-owned industries, China's military industries have had better access to the center and obvious advantages in the supply of energy and raw materials, in the use of transportation, and in raising funds. Consequently, the total output of China's military industries (including aviation) could, and did, increase enormously. This growth in itself has changed the proportion between military and civilian production.

These conclusions are firmly supported by Chinese data (see Table 6.2). Whereas the entire output of China's aviation industry in 1993 was 5.34 times higher than in 1979, civilian production was 23.4 times higher. This has been achieved by activating underutilized civilian production capacity and by expanding it—hardly at the expense of military production. Although the share of military production in total aviation industry production has indeed plunged from about 95 percent in 1978 to about 25 percent in 1995, actual military production has not changed significantly. In fact, from 1979 to 1984 military aviation production increased and in 1984 was 83.7 percent higher than in 1979. Military production indeed began to decline after 1985, but the yearly average for 1979–1988 (2.50 billion yuan) was still higher than in 1979 (2.33 billion yuan). Although no data for military production have been available since the late 1980s, it may be safely assumed that even if military production continued to decline in quantitative terms (it may be the other way around), it has been qualitatively, technologically, and financially upgraded by China's policy of military-civilian combination.[67]

Few military production lines have been converted, and most of them are either too specialized or too outdated to be of much use. In fact, there is a strict separation between military and civilian production lines and this separation increases as conversion proceeds. While underscoring civilian production, China's aviation industry spares no effort in modernizing its military production lines. Domestic orders must have declined considerably because, especially following the troop reductions, the PLAAF is saturated with obsolete aircraft. Yet at the same time international orders have multiplied, keeping some of the more advanced military aviation production lines—using foreign technologies—busy. As shown above, Beijing invests a good deal of human, material, and financial resources in state-of-the-art military aircraft. Although this balanced combination of military and civilian production is not conventional, it typically reflects traditional culture. In this respect it is "conversion with Chinese characteristics."

Notes

1. Richard M. Bueschel, *Communist China's Air Power* (New York: Praeger, 1968), p. 39.

2. Information on the development of China's aviation industry is drawn mainly from the following sources: *Dangdai Zhongguo De Hangkong Gongye* [China Today: Aviation Industry] (Beijing: Zhongguo Shehui Jiaoxue Chubanshe, 1988); Duan Zijun, ed., *China Today: Aviation Industry* (Beijing: China Aviation Industry Press, 1989), hereafter *CTAI; Zhongguo Hangkong Gongye Sishinian (1951.4–1991.4)* [Forty Years of China's Aviation Industry] (Beijing: Hangkong Gongye Chubanshe, 1991); and *China Today: Defence Science and Technology* (Beijing: National Defence Industry Press, 1993), 2 vols., hereafter *CTDST.*

3. More details in *Survey of Chinese Aviation Industry 1993/1994* (Beijing: Aviation Industry Press, 1993); *Shijie Hangkong Gongye Gailan 1988/1989* [Survey of World Aviation Industry 1988/1989] (Beijing: Hangkong Gongye Chubanshe, 1988), pp. 1–32. See also *China Daily,* 24 June 1993, p. 2.

4. *CTAI,* p. 17; Huang Bin and Pan Yanxi, "Flying Freely in Vast Territory and High Sky—Account of Development of New China's Aviation Industry," *Renmin Ribao* [People's Daily, Overseas Edition], 26 September 1994, p. 1, in *Foreign Broadcast Information Service, Daily Report China* (Washington, D.C., hereafter *FBIS-CHI*), 13 October 1994, pp. 62–63.

5. Sergei N. Goncharov, John W. Lewis, and Xue Litai, *Uncertain Partners: Stalin, Mao, and the Korean War* (Stanford: Stanford University Press, 1993), pp. 73–74.

6. *CTAI,* pp. 16–17. For background, see Raymond L. Garthoff, "Sino-Soviet Military Relations, 1945–66," in *Sino-Soviet Military Relations,* ed. Raymond L. Garthoff (New York: Praeger, 1966), pp. 82–99. See also Bueschel, *Air Power,* p. 40.

7. "The Cold War in Asia, Draft Version," *Cold War International History Project Bulletin,* issues 6–7 (Washington, D.C.: Woodrow Wilson International Center for Scholars, winter 1995/1996), p. 11.

8. *CTAI,* p. 20.

9. Ibid., pp. 21–24.

10. Ibid., pp. 50, 54.

11. Ibid., pp. 62–63, 71. For background, see Barry Naughton, "The Third Front: Defense Industrialization in the Chinese Interior," *China Quarterly,* no. 115 (September 1988), pp. 351–386; and Barry Naughton, "Industrial Policy During the Cultural Revolution: Military Preparation, Decentralization, and Leaps Forward," in *New Perspectives on the Cultural Revolution,* eds. William A. Joseph, Christine P.W.

Wong, and David Zweig (Cambridge, Mass.: Harvard University Press, 1991), pp. 153–181, 317–321.

12. "Letter to the staff of the No. 320 Factory (August 1, 1954)," in *The Writings of Mao Zedong 1949–1976,* vol. I, eds. Michael Y. M. Kau and John K. Leung (Armonk, N.Y.: Sharpe, 1986), pp. 466–467. See also *CTAI,* pp. 31–32.

13. John Gittings, *The Role of the Chinese Army* (London: Oxford University Press, 1967), pp. 140–141. See also Bueschel, *Air Power,* pp. 68, 72.

14. Kenneth Munson, "China's Aero Industry Deserves More Credit," *Jane's Defence Weekly,* 14 December 1985, p. 1315.

15. Kenneth Munson, "Fishbed, Finback and the Chinese Future," *Jane's Defence Weekly,* 21 December 1985, p. 1367; Bueschel, *Air Power,* p. 89.

16. *CTAI,* p. 117; Bueschel, *Air Power,* pp. 90–91.

17. John Wilson Lewis and Xue Litai, *China Builds the Bomb* (Stanford: Stanford University Press, 1988), pp. 244–245.

18. *CTAI,* p. 69.

19. Ibid., pp. 175–179.

20. *CTDST,* vols. 1 and 2, pp. 86, 614, 618. See also Philip J. Klass, "Avionics in China: China Moves to Counter Effects of Cultural Revolution," *Aviation Week and Space Technology,* 6 June 1983, pp. 52–64.

21. Bueschel, *Air Power,* pp. 41, 107–216; Richard E. Gillespie, "Aircraft Procurement and Production," *China Business Review* (July-August 1986), p. 40.

22. *CTAI,* p. 176.

23. Ibid., p. 42.

24. Paul Humes Folta, *From Swords to Plowshares? Defense Industry Reform in the PRC* (Boulder, Colo.: Westview Press, 1992), p. 154.

25. *CTAI,* pp. 45, 444.

26. *CTDST,* vol. 2, pp. 617–618.

27. *CTAI,* p. 43.

28. Ibid., p. 71.

29. See, for example, Zhang Hongbiao (vice president, China General Corporation of Aviation Industry), "Exploiting Scientific and Technical Advantage, Strengthening International Cooperation to Promote Greater Development of the Aviation Industry's Transition from Military to Civilian Products," paper presented at an International Conference on the Conversion of China's Military Industries, Beijing, June 1995.

30. *CTDST,* vol. 1, p. 24.

31. *CTAI,* p. 406.

32. Ibid., p. 81.

33. Lin Zongtang, "China's Aviation Industry Marching Forward," *Guoji Hangkong* [International Aviation], no. 1 (January 1992), pp. 10–11, in *Joint Publication Research Service (JPRS), Chinese Science and Technology* (Washington, D.C.), 16 April 1992, pp. 9–10.

34. Zhu Yuli, "China's Aviation Industry Spreads Its Wings," *Beijing Review,* no. 13 (1994), pp. 8–11.

35. *CTAI,* pp. 449–452.

36. Ibid., pp. 80, 84.

37. Ibid., p. 444.

38. Ibid., p. 76.

39. Xinhua news agency, 26 August 1993, in *FBIS-CHI,* 27 August 1993, pp. 38–39; Qiu Qi, "Converted Factories Export $560 Million," *China Daily–Business Weekly,* 20–26 November 1994, p. 1.

40. *CTDST,* vol. 1, p. 147.

41. *CTAI,* p. 448. For a brief overview see Cao Shixin, ed., *Zhongguo Junzhuanmin* [Chinese Defense Conversion] (Beijing: Zhongguo Jingji Chubanshe, 1994), pp. 66–78; Xu Xinglin and Gao Xiaoxing, "The Conversion Experience of the Chinese Aviation Industry," in *Restructuring the Military Industry: Conversion for the Development of the Civilian Economy,* ed. Qian Haiyan (n.p.: CAPUMIT and United Nations Department of Development Support and Management Services, 1994), pp. 261–266.

42. *CTAI,* p. 173. See also Klass, "Avionics in China," p. 53.

43. "China's Aviation Industry Flies High," *Beijing Review,* 30 January 1995, p. 28.

44. Zhu, "China's Aviation Industry," pp. 8–11. See also Gao Jinan, "Aero-Industry to Take Off on Long Flight," *China Daily,* 24 June 1993, p. 2; Ding Xuemei, "Nation Launches New Era in Aviation Development," *China Daily–Business Weekly,* 2–8 October 1994, p. 8.

45. Xinhua, 4 January 1995, in *FBIS-CHI,* 4 January 1995, pp. 46–47; Xinhua, 21 January 1995, in *FBIS-CHI,* 23 January 1995, p. 45; Xinhua, 23 January 1995, in *FBIS-CHI,* 25 January 1995, p. 47. See also Adam M. Pilarski, "Forecasting China's Aircraft Needs," *China Business Review* (September-October 1987), pp. 34–35; Beverly LoPinto, "Airport Development," *China Business Review* (July-August 1986), pp. 43–48; Joseph W. Lee. "The Sky's the Limit: China's Overburdened Aviation Sector Seeks U.S. Know-How," *China Business Review* (May-June 1993), pp. 12–18.

46. Data adapted from several sources, including *Shijie Hangkong Gongye Gailan 1988/1989* [Survey of World Aviation Industry] (Beijing: Hangkong Gongye Chubanshe, 1988), pp. 1–32; Jin Zhude, ed., *Guide to International Corporation and Investment with Enterprises of China's Defense Industry,* 2 vols. (Beijing: China Association for Peaceful Use of Military Industrial Technology, 1993); *Handbook of Chinese Aviation Industry Trade Group* (Beijing: Aviation Industries of China, 1993); *Survey of Chinese Aviation Industry 1993/1994* (Beijing: Aviation Industry Press, 1993).

47. For a summary of China's relations with MD, see Madelyn C. Ross, "McDonnell Douglas: The Management Challenge," *China Business Review* (September-October 1987), pp. 36–38.

48. *Wen Wei Po* (Hong Kong), 3 July 1986, p. 1, in *JPRS–Chinese Economic Affairs,* 22 October 1986, p. 81; Xinhua, 19 September 1994, in *FBIS-CHI,* 20 September 1994, p. 36; Sun Shang, "Country's Aviation Flying High with Foreign Firms," *China Daily,* 23 January 1995, p. 1.

49. Zhang Yuan, "Airplane Maker Keen on Foreign Partners," *China Daily–Business Weekly,* 26 September 1993, p. 6.

50. Liang Fang, "Guizhou Arms Industry Converts Production," *Beijing Review,* no. 41 (11–17 October 1993), pp. 12–15.

51. *CTAI,* p. 180.

52. Jiangxi Radio, 13 November 1994, and Xinhua, 16 November 1994, in *FBIS-CHI,* 17 November 1994, pp. 30–31; Jiangxi Radio, 16 November 1994, in *FBIS-CHI,* 23 November 1994, p. 60.

53. Xinhua, 24 March 1993, in *FBIS-CHI,* 12 April 1993, pp. 28–29; Xinhua, 30 March 1995, in *FBIS-CHI,* 30 March 1995, pp. 45–46; Xinhua, 3 April 1995, in *FBIS-CHI,* 3 April 1995, pp. 57–58. For more details on HAMC, see also Paul Proctor, "Harbin Uses New Helicopter Program to Advance Global Manufacturing Role," *Aviation Week and Space Technology,* 3 February 1992, pp. 48–50.

54. Gillespie, "Aircraft Procurement," p. 40. For more details on XAC, see Paul Proctor, "China's Xian Aircraft Corp. Emphasizes Y-7 Production, Foreign

Subcontracts," *Aviation Week and Space Technology,* 24 February 1992, pp. 112–117; *Xi'an Aircraft Company* (Xi'an: Xi'an Aircraft Company Overseas Production, 1994).

55. Xinhua, 27 January 1995, in *FBIS-CHI,* 30 January 1995, p. 46.

56. Xinhua, 2 September 1994, in *FBIS-CHI,* 2 September 1994, p. 13. See also Gao Jinan, "Aerospace Industry's Peaceful Products," *China Daily,* 29 January 1993, p. 1; Xinhua, 14 February 1993, in *FBIS-CHI,* 17 February 1993, p. 31; Xinhua, 19 September 1994, in *FBIS-CHI,* 20 September 1993, p. 36; Qiu, "Converted Factories," p. 1.

57. Zhu, "China's Aviation Industry," pp. 8–11.

58. *Zhongguo Xinwen She,* 12 September 1993, in *FBIS-CHI,* 23 September 1993, pp. 39–40; Xinhua, 28 October 1994, in *FBIS-CHI,* 1 November 1994, p. 61.

59. Xinhua, 2 April 1994, in *FBIS-CHI,* 4 April 1994, p. 47.

60. Xiao Yong, "Aviation Industry Counting on More Loans," *China Daily–Business Weekly,* 2–8 October 1994, p. 1.

61. *CTDST,* vol. 1, pp. 118–119.

62. R. Bates Gill, *Chinese Arms Transfers: Purposes, Patterns, and Prospects in the New World Order* (Westport, Conn.: Praeger, 1992), pp. 217–218. See also: "The Sino-Italian A-5M," *International Defense Review,* no. 8 (1987), p. 1050.

63. *Kyodo,* 8 January 1993, in *FBIS-CHI,* 8 January 1993, p. 9; *Yomiuri Shimbun,* 9 November 1993, p. 1, in *FBIS-CHI (Annex),* 15 November 1993, p. 6; *AFP,* 19 January 1994, in *FBIS-CHI,* 25 January 1994, pp. 55–56; *Nihon Keizai Shimbun,* 28 January 1994, p. 8, in *FBIS-CHI (Annex),* 4 March 1994, p. 10; *Yomiuri Shimbun,* 8 January 1995, p. 4, in *FBIS-CHI,* 10 January 1995, p. 34.

64. Stockholm International Peace Research Institute, *SIPRI Yearbook 1993: World Armaments and Disarmament* (Oxford: Oxford University Press, 1993), pp. 498–511.

65. Ma Chenguang, "Exports of Jet Trainers Take Wing," *China Daily,* 22 September 1994, p. 1. The K-8 was designed to U.S. specifications and incorporates U.S. equipment, including a Garrett TFE731-2A turbofan engine; Michael A. Dornheim, "Karakoram K-8 Trainer Makes Air Show Debut," *Aviation Week and Space Technology,* 9 March 1992, pp. 42–44.

66. On some of these problems, see Yitzhak Shichor, *China's Defence Capability: The Implications of Military-to-Civilian Conversion,* CAPS Papers No. 8 (Taipei: Chinese Council of Advanced Policy Studies, April 1995); Yitzhak Shichor, *Military-to-Civilian Conversion in China: From the 1980s to the 1990s,* Working Paper No. 142 (Canberra: Peace Research Centre, Research School of Pacific Studies, Australian National University, December 1993).

67. Folta, *From Swords to Plowshares?* pp. 166–168. For a more quantitative and lenient approach to Chinese conversion, see Jean-Claude Berthélemy and Saadet Deger, *Conversion of Military Industries to Civilian Production in China: Prospects, Problems and Policies* (OECD Development Center Report, draft, May 1995), p. 5.

7

The Regional Implications of Defense Conversion: The Case of the "Third Line" and Guizhou

Arthur S. Ding

The large-scale conversion of the military industry to civilian production in the People's Republic of China (PRC) has had a major impact on patterns of Chinese economic development. In this effort, Beijing has been forced to confront the problems of the "Third Line" (*sanxian*) industries—that major and expensive portion of the defense sector located in China's remote interior. The Third Line itself has at least two levels: the national sector under central control, which has concentrated on the development of major equipment such as tanks, aircraft, warships, and heavy artillery for the regular armed forces; and "little Third Lines" (*xiao sanxian*), which manufactured rifles and ammunition for use by the militia.[1]

Since the beginning of the reforms, however, Beijing has focused its support on the national sector and has neglected the *xiao sanxian*. Due to the small scale of the *xiao sanxian* and their simple manufacturing processes, local governments have found it easy to influence what they produce, and they have been quickly integrated into the local economy. Yet gaps still remain between the defense and civilian sectors.

Problems and Assumptions

The study of these developments in the provinces requires that we make a number of assumptions to compensate for gaps in the data. Although a reasonably complete data set on the defense sector can be assembled from the economic yearbooks of Guizhou and there are useful data from Sichuan, most provinces hide defense industry data within their general statistics. At the same time, most new, post-1979 industrial investment in the Third Line regions has been directed at correcting the economic imbalances and distortions created by Third Line projects, and we think it safe to assume that much, if not most, of this investment has gone into conversion.

Structural Issues

The problems in the interior arise not only from the massive investment in defense projects at the expense of local development but also from the economic structures that governed the Third Line.[2] These defense enterprises operated in a highly centralized environment that was based on the Soviet model, in which central ministries made all the decisions concerning personnel, new departments, new product development, and allocation of capital. The enterprises were simply agents of central decisions and had no links with the external economic environment. Production was carried out according to administrative orders—there were no incentives and, in this top-down system, there was no room for feedback from the bottom.

These ministries in turn had a high degree of vertical integration, which resulted in a closed, self-contained management system, with each ministry operating its own autonomous system of enterprises. Enterprises in one system had no access to the facilities of those in other systems, resulting in duplication of investment, equipment, and effort.

In the end, the *sanxian* was separated—indeed, isolated—from the civilian economy. The huge amount of central investment was made without regard for the needs of the civilian sector or the overall deployment of resources. There was no contact between and no transfer of skills or resources from the defense to the civilian economy. Particularly in the provinces, a dual economy of high-technology defense plants operating in a low-technology civilian environment was created. Moreover, this system perpetuated itself: When the time came for defense enterprises to find partners for cooperative projects, they were forced to look to the more developed coastal regions or to other defense enterprises because the local civilian sector was insufficiently advanced.

Worse yet, although the defense sector contributed little to the local economy, local governments shouldered vast obligations in support of these enterprises. Local governments had to provide housing, food, water, power, and other infrastructure needs while receiving very little benefit in return. For example, the Sichuan provincial government had to subsidize its defense industries by 72 million yuan annually—yet received only 50 million yuan in revenue from those enterprises.[3]

All of this had a detrimental effect on regional development. Although statistics from the provinces in the prereform period indicate an overall improvement in economic performance, these are only averages for each locality. It appears that defense enterprises operating within the national plan still do better than the local civilian sector.

Reform

The economic reforms launched in 1979 attacked many of these structural problems, which affected both the civilian and defense sectors. Without these reforms, a list of which follows, conversion to civilian production would have been impossible.[4]

- *Contract system.* Probably the most important of all the structural reforms, contracts would allow the system to open up because enterprises could now form alliances and approach the market with extra products. Contracts—along with the *factory manager responsibility* system, in which managers were given profit-and-loss responsibilities—increased enterprise autonomy; the system was soon expanded to include defense research institutes. This new freedom, however, did not mean that factories or institutes could reject tasks specified by the state.
- *Price and financial reform.* Enterprises were allowed to fix their own prices for goods produced outside the national plan—including conversion products. Profits from the sale of those goods could be retained. This, in turn, encouraged the development of new accounting systems. Financing shifted from allocations to the use of local government revolving funds, bank loans, and equity placements. Priority research and weapons production, however, remained funded by the state, and the prices of weapons were held down by controls that limited them to 5 percent over the cost of production.
- *Export opportunities.* Similarly, defense industrial enterprises were allowed to export and to retain a proportion of their foreign exchange earnings, a proportion actually larger than that allowed civilian enterprises. At the same time, however, enterprises were encouraged to avoid excessive imports, particularly of technology that was available in China; if enterprises imported foreign equipment, it was to be reverse-engineered. The aim was to increase the competitiveness of their civilian products; profits from these could then be used to facilitate the production of modern weapons.
- *Labor reform.* Through a system of contracts between enterprises and workers, pay was linked to performance, training became available, and there was some increase in labor mobility, even for technicians.

Special Assistance

In addition to the these four reforms, which also applied to ordinary civilian enterprises, the government offered special assistance to the defense industrial sector—especially in the Third Line regions—to facilitate conversion to civilian production. One important measure was to relocate enterprises and factories from remote mountainous areas to urban areas of the heartland or even to the coastal provinces. The purpose of this was to bring these enterprises closer to their markets and to improve their communication with the outside world, allowing them to integrate into the civilian economy. The PRC's international situation was no longer as tense as it had been at the beginning of the 1960s when the Third Line enterprises were established; they could now be moved. The government drew up a number of regulations to facilitate relocation. One example was the resolution passed at the Third Line adjustment work conference held in August 1984, which called for the transfer of defense enterprises scattered in remote mountainous areas or areas with problematic geological structures.[5] The transfer of these enterprises was included in the national plan, and the costs of moving machinery and equipment and providing new housing for employees and schooling for their children were to be assumed by the central and local governments.

There is little available information concerning how much the government has invested in relocating Third Line enterprises. The scarce existing data indicate that in the period of the Seventh Five-Year Plan (1986–1990), 2 billion yuan was spent on relocation and reorganization—800 million yuan of which was provided by the state in the form of direct loans and loans from development banks. The ministries in charge of the enterprises invested another 600 million yuan, and the enterprises themselves provided the remaining 600 million yuan. This investment was spread among 121 production enterprises and research units, all in imminent danger of collapse, with a total workforce of over 160,000. The relocation of these enterprises was expected to be completed in 1990.[6]

Guizhou Moves

Guizhou is one of the provinces that has benefited from the 1986–1990 relocation drive, with twenty-four enterprises and institutes included in the program at a cost of 381 million yuan. Of these, fifteen factories were moved to Guiyang, five factories were moved to Zunyi, and one research institute was moved to Hefei in Anhui province. Another three factories remained in their original locations, but their product ranges were adjusted.[7] By 1990, twelve factories had been relocated, one had completed product adjustment, and the others had completed basic engineering work in preparation for removal.[8]

The biggest beneficiary of the relocation program in the province seems to have been the Guizhou Aviation Industrial Corporation. By the end of 1992, one enterprise belonging to this corporation had undergone transformation, while eight had completed removal of equipment, materials, and 14,500 employees and their families. The enterprises, which were formerly scattered over a large area, are now concentrated in Anshun, Pingshun, and Guiyang.[9]

Civilian Products and Technology

Another form of assistance offered by the government is help with the selection of civilian products. The government encourages defense enterprises to select products of an appropriate technological level in order to take advantage of their technology and workforce. But to manufacture civilian products, enterprises—particularly those involved in mechanical or electrical engineering—often must purchase new equipment. Furthermore, existing equipment is often old and outdated; even when equipment can be used to manufacture civilian products, quality tends to be low and products lack competitiveness. Therefore, defense industrial enterprises need to invest large amounts of capital into the purchase of new equipment.

The government, however, is ready to help these enterprises carry out the technological renovation necessary for converting to civilian production. To do this, the government first made provisions for technological renovation in the national plan and then provided capital grants or loans for the purchase of new equipment or technology.

The government has poured large amounts of money into technological renovation. During the Seventh Five-Year Plan period (1986–1990), loans worth 4 billion yuan were granted to renovate 296 items of technology and to construct 450 production lines for civilian products. During the Eighth Five-Year Plan (1991–1995), 6.3 billion yuan in loans was spent on the renovation of more than four hundred items of technology.[10]

A number of enterprises in Sichuan province have benefited from the technological renovation policy. During the Sixth Five-Year Plan (1981–1985), 785 million yuan was invested into the technological renovation of defense industries in the province—1.59 times the amount invested in production-related capital investment. Of this amount, 57 percent was spent on reducing energy consumption, raising the quality of products, increasing production volume, and expanding the range of products.[11]

A similar kind of assistance is available for capital construction, particularly for major items included in the national plan. For example, the government intends small cars to be one of the main items produced by Third Line enterprises after their conversion to civilian production. To ensure efficiency, the Beijing authorities are making arrangements in the national plan and providing capital to enable the provinces concerned to construct automobile manufacturing zones.

The provincial government has been particularly helpful in terms of capital construction in Guizhou. In 1986 the province found sites, acquired land, and approved plans for enterprises and factories to be relocated from remote areas. In addition, the Guiyang city government granted privileged treatment to Third Line enterprises that moved into the city, giving them money to acquire land, exempting them from urban facilities fees, granting their employees residence in the city, and providing various tax reductions and exemptions.[12]

Enterprises under the Ministry of Astronautics made full use of these privileges. The ministry had decided that Guizhou was to be an important center for manufacturing compact automobiles.[13] Beginning in May 1988, the ministry decided to invest a substantial amount of funds in developing Zunyi into a light automobile manufacturing city; to do this, enterprises were moved into the area and simultaneously subjected to technological renovation. By the end of 1992, 188 million yuan in central government funds had been spent on this project, 83.6 percent of the total amount of investment allocated by the ministry. The entire site occupied 290,000 square meters, 170,000 square meters of which had already been completed. The project included fifteen items of technology renovation, at a total cost of 290 million yuan. Purchases included steel pressing equipment; welding, coating, and heat treatment equipment; gear boxes; and molds.[14]

Marketing

The government also assists military industries in marketing their civilian products. It was noted previously that friction between military industrial enterprises and local governments makes it difficult for the products of the former to compete on the local market, and this hinders their development and threatens their survival. The government has adopted a number of different solutions to this problem. One is to gradually devolve authority over some military industrial enterprises to local governments. Indeed, control over military electronics enterprises has been devolved entirely to the local level.[15]

The China Zhenhua Electronics Industrial Company in Guizhou province is one example of an enterprise that has been handed over to local control. In 1984, the Ministry of Electronics Industry reorganized its thirty-one units in Guizhou into the China Zhenhua Electronics Industrial Company, centered on what was formerly the ministry's Guizhou office. The new company exercised unified management over personnel affairs, finance, materials, production, supplies, and marketing. In November 1985, the China Zhenhua Company was placed under the joint control of the Ministry of Electronics Industry and the Guizhou provincial government.[16] In August 1986, the provincial government established an electronics industry office to take charge of both the civilian and the military electronics industry. This initiative dissolved the barriers between the vertical and horizontal chains of command and between the military and the civilian sectors.[17]

Similar adjustments have been made among industries under the Ministry of Aviation Industry. In March 1986, this ministry announced that its office in Guizhou was an agency of the ministry under the joint leadership of the ministry and the Guizhou provincial government. This office has a contract with the ministry to exercise unified control over all aeronautics industry enterprises in the province.[18]

Sometimes defense enterprises that have been devolved to the provincial level are included in the province's annual economic plan. In a planned economy, the local administration is obliged to provide all necessary production resources to enterprises within the plan. In this way, the central government both reduces its own burden by passing it on to local governments and increases local rights and obligations. For the enterprises, this solution has the advantage of bringing their products into the local marketing system and allowing them to become further integrated into the local economy, thus reducing the possibility of friction with local authorities.

In 1990, the central government made a new pronouncement concerning the relationship among local governments, defense industrial enterprises, and the economic plan: "The policy of military–civilian integration is not simply the concern of the departments of science, technology and industry for national defense, it is the joint responsibility of all departments and all localities."[19] Under the influence of this pronouncement, Guizhou province issued a "Decision on the Gradual Strengthening of Military-Civilian Integration Work," which contained the following proposals: (1) science, technology, and industry for national defense should be integrated with the local economy; (2) local government departments should actively support and accurately guide military industrial enterprises in carrying out structural adjustments, helping them choose good production projects and drawing up plans for developing civilian products; and (3) local government should enhance its supervision over and coordination of plans for the production of civilian products by military industrial enterprises, and important civilian products should be included in the national, ministerial, and provincial plans.[20]

This kind of support is very important for enterprises just embarking upon conversion to civilian production. These enterprises are entering a completely new field, and it is especially difficult for them to win a share of the market when they are producing technologically advanced goods such as automobiles. They have not yet reached economies of scale and their costs are high; to survive they must rely on administrative assistance to find markets for their products. For example, the number of small sedans produced by automobile plants under the Ministry of Aeronautics and Astronautics was zero in 1991, five in 1992, and 520 in 1993; the corresponding figures for minibuses were zero, 1,659, and 1,999.[21] These products were all sold through the local government's marketing system.

Local Links and Preferential Treatment

The central government now attaches great importance to cooperation between defense industrial enterprises and local authorities, particularly to the effects of such cooperation on conversion to civilian production.

> We know from experience that when local leaders pay attention [to an enterprise] and include its civilian products in the local economic development plan, when military industrial departments actively rely on the localities, and when important problems can be promptly discussed with the local government, military-civilian integration is more successful. At present, enthusiasm for the development of civilian products by military industrial enterprises is low in some localities; this is a structural problem and a problem of economic interests.[22]

Inclusion of defense industrial enterprises in local planning systems solves both of these problems at a single stroke.

The central government also assists defense industrial enterprises by giving them preferential treatment in granting contracts for major development projects. A typical example is the role played by the China General Corporation of the Nuclear Industry in the design and construction of the Qinshan Nuclear Power Plant in Zhejiang province. This kind of assistance is similar in nature to the inclusion of civilian products manufactured by defense industrial enterprises in the local planning system, except that in this case they are included in the central government's plan.

Reorganization

In addition to cooperation with local entities, another solution to the problems afflicting defense industrial enterprises is reorganization. In the prereform period, the chain of command in the defense industry was strictly top-down. Individual enterprises and plants did no more than mechanically carry out the production tasks assigned them by their superior units. Each individual industry was self-contained and self-sufficient; barring instructions to the contrary from above, there was no contact or cooperation of any kind between units belonging to different "systems." This setup had an extremely detrimental effect on economic development.

The government's solution to this problem was reorganization of the defense industry, particularly by the formation of enterprise conglomerates (*jituan*). In China a conglomerate consists of any group of companies linked by some form of cooperation—for example, a core plant and a number of satellite plants—rather than the collection of enterprises linked by a holding company found in the West. In addition, the meaning of the term "conglomerate" has expanded since the beginning of the reforms.[23]

The formation of enterprise conglomerates has chiefly served to break down the vertically organized, self-contained industrial systems of the past. The authorities hope that the formation of conglomerates—especially those that bring together technological development, industry, and trade and encourage division of labor, economies of scale, and mutual assistance—will promote overall development. Enterprise conglomerates have a fairly large degree of autonomy, and they undertake projects under contract from their superior units.[24] The development and manufacture of military products, however, are still controlled by the central plan. Although these enterprise conglomerates are based in the provinces, they remain the property of their superior units; taxes and profits are handed over to these units, not to the local governments.[25]

The defense industrial enterprises of Guizhou province are among those that have undergone such reorganization, which commenced in 1986 when it was decreed that enterprises under the Ministry of Aviation Industry should be administered locally. The second wave of reorganization came two years later, in April 1988, when two companies were established: the Guizhou General Corporation of the Aviation Industry, consisting of forty-six enterprises with registered capital of 1.5 billion yuan; and the Guizhou General Corporation of the Astronautics Industry, with thirty-eight enterprises and registered capital of 1.2 billion yuan.[26]

Military-industrial conglomerates made their first appearance in Guizhou in 1992. There were three main conglomerates: the Guizhou Aviation Industrial Conglomerate, whose core enterprise was the Guizhou General Corporation of the Aviation Industry; the Guizhou Astronautics Industrial Conglomerate, centered on the Guizhou General Corporation of the Astronautics Industry; and the Zhenhua Electronics Conglomerate, based on the China Zhenhua Electronics Industrial Corporation. These were three among the first batch of fifty-five experimental enterprise conglomerates set up by the government.[27]

Enterprise conglomerates have certain privileges. For example, the Guizhou Aviation Industrial Conglomerate—consisting of forty-four closely connected units, twenty-two somewhat closely connected units, and more than two hundred loosely connected units—is allowed to import and export and to make decisions concerning external cooperation in funding and technology.[28]

The formation of enterprise conglomerates promotes production and economic linkages between defense industrial enterprises and their local counterparts. For example, in the production of compact automobiles, the main item produced by the Guizhou Astronautics Industrial Conglomerate, the core company provides its local satellite plants with technology, equipment, and management expertise to assist them in producing more than 170 different automobile parts, including windshields and ignition coils. Moreover, the company has undertaken joint investments worth 110 million yuan with local governments—including township enterprises in the automobile production zones—thus expanding the production of automobile parts and other related products.[29] These are the relationships described as loose or somewhat close. Trends such as these increase inter- and intra-regional economic interaction.

Negative Factors Influencing Conversion in the Third Line Regions

The measures discussed above have indeed been helpful in promoting economic development in the Third Line regions. For example, the selection of appropriate civilian products such as automobiles; investment in technological upgrading, capital construction, and relocation; and the establishment of regional systems for the division of labor have been particularly effective. To some extent, recent investment recalls that of the past, when the government was willing to pour in large amounts of capital without requiring any improvement in efficiency. This time, however, managerial reforms have made a contribution to economic growth.

Nevertheless, conversion to civilian production in the Third Line regions has not taken place in a vacuum; it is part and parcel of a broader process of economic reform rather than a new version of the mid-1960s policy of concentrating national resources in Third Line defense industries. The reforms have not been as beneficial to the Third Line defense industries or to economic growth in the regions in which they are located as might have been expected. Although conversion to civilian production has breathed new life into the local economies in these areas, they have not done as well as other parts of the country.

Central government investment in the Third Line regions has not been sufficient to have an impact on economic growth. According to conventional wisdom, the main problem is still a shortage of capital. There are several reasons for this shortage. First, the state is not honoring its economic subsidy commitments. As a result of financial troubles, by 1990 the government had supplied only 77.2 percent of the subsidies it had promised the Third Line regions under the Seventh Five-Year Plan. Second, inflation has driven up capital construction costs. Third, ministries and enterprises cannot obtain the small proportion of capital they are required to raise themselves, leaving them in a situation in which they can neither continue operating in their original plants nor afford to complete the new ones. Fourth, the new plants provided for relocated enterprises are often too big, exacerbating the problem of insufficient capital. Finally, the government failed to honor its commitment to provide certain privileges to these industries, such as interest payment subsidies and tax reductions and exemptions. In particular, financial belt-tightening forced the government to reduce its subsidies on interest payments.[30]

More than the lack of capital or the government's failure to honor its promises, the entire trend of economic reform has had the most marked detrimental effects on conversion to civilian production in the Third Line regions. The economic reforms encouraged a redistribution of resources that favors the southeastern coastal region of China rather than the Third Line provinces as in the past. This trend began during the period of the Fifth Five-Year Plan (1975–1980) and accelerated in the 1980s. As a result, the Third Line regions have not been able to obtain the development capital they require. For example, in 1992, investment in fixed assets by state-owned units throughout the country totaled 758.15 billion yuan, of which Guizhou secured only 6.59 billion yuan.[31] The situation was just as bad the following year, when out of a total investment of 832.12 billion yuan only 7.29 billion yuan went to Guizhou.[32] This reduction in the supply of capital deprived the Third Line regions of much-needed support, and to some extent canceled out the benefits of conversion to civilian production carried out by their defense industries.

Not only are the Third Line regions not getting enough financial support from the state, but they are also falling even further behind the coastal regions in terms of foreign investment. In 1992, the PRC attracted more than 1.1 trillion yuan in foreign direct investment (not counting foreign loans), but less than 2 billion yuan of that amount—less than 1 percent of the total—was invested in Guizhou province.[33] In 1993, foreign direct investment exceeded 2.7 trillion yuan, but only 4.3 billion yuan found its way to Guizhou, once again less than 1 percent of the total.[34] While Shanghai, Guangdong, Zhejiang, and Fujian are all attracting significant quantities of foreign capital, Guizhou has been overlooked by foreign investors. Consequently, the conversion to civilian production undertaken by its defense industries cannot be expected to bring a substantial increase in economic growth. Furthermore, as the Third Line regions are primarily engaged in capital-intensive heavy industries and technology-intensive production, the shortage of capital may have fatal consequences.

Another problem is that much of the investment undertaken in the Third Line regions since the beginning of the reforms has not been profitable, particularly investment devoted to relocated plants situated in remote areas. These plants were on the brink of collapse, and in relocating them the authorities were chiefly motivated by political and social rather than economic considerations. The government was afraid that their collapse would throw enormous numbers of people out of work, creating massive social and political problems. Investments carried out for these reasons could hardly be considered productive in nature or able to promote economic growth; they would not create any additional employment opportunities.[35]

Conversion to civilian production in the Third Line regions also did not have the expected import substitution effect. Many of these plants are equipped with outdated machinery, and it is difficult for them to manufacture the large quantities of high-tech civilian products needed for import substitution. In addition, conversion to civilian production necessitated the import of large amounts of light processing equipment, leaving existing production capacity idle. Because the factories are too distant from markets, the products manufactured with this imported equipment do not sell well and therefore do not serve the purpose of promoting economic growth.

Finally, although Third Line defense industries have obtained a limited amount of capital from the state and a small amount of foreign direct investment, they have experienced quite a serious outflow of capital, and this outflow is unlikely to produce any rewards for the enterprises themselves. The channels through which this capital leaves the region are the "windows" set up by Third Line enterprises in the coastal provinces in an effort to obtain information, markets, managerial expertise, and technology. Since 1990, approximately eight hundred Third Line enterprises have opened more than one thousand windows and branches in the coastal provinces.[36] The defense industries of Guizhou are no exception: In 1986, Guizhou defense enterprises set up 101 windows, although they had established only seventy-five linkages with other enterprises in their home province.[37] Although such figures do not mean very much in themselves, they do indicate that many Third Line defense enterprises are more interested in establishing links with the coastal provinces than in investing locally.

The real problem, however, is not simply the existence of the windows but rather the form they take. Third Line enterprises are state-owned firms under ministerial supervision, whereas the windows are private enterprises in which the Third Line enterprises have invested and perhaps in which some of their employees hold shares. Ministries cannot exercise supervision over them; they are entirely profit-oriented. It is interesting to note that in some cases, the president of the window company is also the president of the parent Third Line enterprise—for example, the president of the Zhenhua Electronics Company in Shenzhen is also the president of the China Zhenhua Electronics Industrial Conglomerate in Guizhou.[38] This rules out effective supervision, and the pursuit of self-interest makes it impossible for any return on this capital outlay to make its way back to Guizhou.[39]

In an effort to reap some return on its investment, the Zhenhua Company has drawn up new regulations to ensure that its windows promote development in the interior. These include the establishment of an economic development department in the Shenzhen company to organize the transfer of technology and capital to the Guizhou enterprise and the development of assistance schemes to help the parent enterprise such as employing workers from the Guizhou enterprise on three-year contracts.[40] However, none of these measures can change the fact that capital, technology, and workers tend to remain in the coastal areas and do not return to the interior.

Conclusion

More than ten years have passed since China decided to restructure the defense industries of the Third Line area. This decade may be characterized as a period of laying the foundation for conversion adjustment, exemplified by the movement of factories from deep mountain areas into newly built industrial zones of vicinity cities and by the establishment of new civilian product lines with state loans and self-raised funds in line with state planning.

It cannot be denied that such measures have helped somewhat to promote regional economic development in the Third Line area. They are similar to those adopted in the prereform period, which relied heavily upon continuous state funding rather than improved efficiency of economic units.

However, there are limitations on the state-sponsored development formula for the Third Line area in the postreform period. Initially, Beijing had placed priority on the coastal areas, according to the notion that one portion of the nation should gain wealth before the other portion. The Third Line area faced tough competition from the beginning, including the coastal area's comparatively easy access to foreign funds, foreign markets, information, and other advantages. At the same time, civilian products of the Third Line area were less competitive in China's domestic market than those of the coastal area.

It therefore seems safe to conclude that the contributions of defense conversion to development in the Third Line area will be eroded by inherently unfavorable conditions there. Possible consequences include continuous high unemployment rates and outflow of migrant workers to the coastal area, and disparities will widen further as the Third Line area remains peripheral and secondary to the coastal area.

Notes

1. For details on the Third Line (or Third Front), see Barry Naughton, "The Third Front: Defense Industrialization in the Chinese Interior," *China Quarterly,* no. 115 (September 1988), pp. 351–386; Jean-Claude Berthélemy and Saadet Deger, *Conversion of Military Industries to Civilian Production in China: Prospects, Problems and Policies* (OECD Development Center Report, final draft, July 1995), pp. 55–82; John Wilson Lewis and Xue Litai, *China's Strategic Seapower: The Politics of Force Modernization in the Nuclear Age* (Stanford: Stanford University Press, 1994), pp. 88–102. Third Line enterprises were established in Shaanxi, Sichuan, Yunnan, Guizhou, and Hunan provinces, and in the western part of Hubei province. Similar concentrations of defense industries in the coastal provinces were known as "little Third Lines" (*xiao sanxian*). Beijing's *xiao sanxian* was located in suburban Changping county, whereas Shanghai's was in eastern Anhui province. Between 1965 and 1978, the central government's investment in Third Line industries in Sichuan province accounted for 60 percent of total state investment in the province since 1949. See *Sichuan Jingji Nianjian* 1986 [Sichuan Economic Yearbook 1986] (Chengdu: Sichuan Shehui Kexue Chubanshe, December 1986), p. 254. During the period of the Third Five-Year Plan (1966–1970), investment in Third Line industries accounted for 52.7 percent of total state investment, and dur-

ing the Fourth Five-Year Plan it accounted for 41.1 percent (or 44.3 percent) of the total. See Naughton, "The Third Front," p. 365.

2. See the discussion of the impact of this system in You Qianzhi et al., eds., *Zhongguo Guofang Jingji Yunxing Fenxi* [An Analysis of the Workings of China's Defense Economics] (Beijing: Zhongguo Caizheng Jingji Chubanshe, February 1991), pp. 105–106.

3. Ibid., p. 108.

4. The main source for these reforms is Paul Humes Folta, *From Swords to Plowshares? Defense Industry Reform in the PRC* (Boulder, Colo.: Westview Press, 1992), pp. 81–109.

5. At the end of 1983, the Beijing government set up an Office for Third Line Adjustment and Reform under the State Council, to provide systematic assistance to the Third Line areas in coping with postreform problems.

6. Sichuan Shehui Kexueyuan Sanxian Ketizu [Third Line Task Force, Sichuan Academy of Social Sciences], "Woguo Sanxian Gongye Zhengce De Tiaozheng" [The Adjustment of Our Country's Policy Toward Third Line Industries], in *Kaifa Da Zinan, Diqu Chanye Juan* [Developing the Great Southwest, Regional Industries] (Beijing: Xueyuan Chubanshe, September 1991), p. 507.

7. *Guizhou Nianjian* 1987 [Guizhou Yearbook 1987] (Guiyang: Guizhou Renmin Chubanshe, November 1987), p. 337.

8. *Guizhou Nianjian* 1993 (Guiyang: Guizhou Renmin Chubanshe, August 1993), p. 345.

9. Ibid.

10. Cao Shixin, ed., *Zhongguo Junzhuanmin* [Chinese Defense Conversion] (Beijing: Zhongguo Jingji Chubanshe, 1994), p. 15.

11. *Sichuan Jingji Nianjian* 1987 (Chengdu: Sichuan Shehui Kexueyuan, December 1987), p. 257.

12. *Guizhou Nianjian* 1987, pp. 337–338.

13. Ibid., p. 484; *Guizhou Nianjian* 1989 (Guiyang: Guizhou Renmin Chubanshe, September 1989), p. 388. Compact automobiles are those with engines under 1,200 cubic centimeters.

14. *Guizhou Nianjian* 1993, pp. 347–348.

15. Interview with PRC official, 22 June 1995.

16. *Guizhou Nianjian* 1986 (Guiyang: Guizhou Renmin Chubanshe, 1986), p. 442.

17. *Guizhou Nianjian* 1987, p. 491.

18. Ibid., p. 483.

19. Cao, p. 5.

20. *Guizhou Nianjian* 1992 (Guiyang: Guizhou Renmin Chubanshe, August 1992), p. 335.

21. *Guizhou Nianjian* 1992–1994 pp. 334, 336, and 341.

22. Cao, p. 21.

23. For an analysis of the Chinese concept of an enterprise conglomerate, see Lou Renhe and Liu Yuzhen, "Jundui Qiye Zujian Qiye Jituan De Jiben Silu" [The Basic Thinking Behind the Establishment of Enterprise Conglomerates by Military Enterprises], *Junshi Jingji Yanjiu* [Defense Economics Studies], no. 12 (1992), pp. 53–56.

24. Ibid.

25. Cao, p. 48.

26. *Guizhou Nianjian* 1989 (Guiyang: Guizhou Renmin Chubanshe, September 1989), p. 387.

27. *Guizhou Nianjian* 1992, p. 335.

28. *Guizhou Nianjian* 1993, p. 345.

29. *Guizhou Nianjian* 1994, p. 341.

30. Li Xiwen and Zhang Ping, "Dui 'Sanxian' Jianshe Tiaozheng Gaizao De Yidian Zhenshi" [Some Understanding of the Adjustment and Reform of "Third Line" Construction], in *Kaifa Da Xinan, Diqu Chanye Juan,* pp. 527–528.

31. *Zhongguo Tongji Zhaiyao* 1993 [Summary of Chinese Statistics 1993] (Beijing: Zhongguo Tongji Chubanshe, May 1993), p. 22.

32. *Zhongguo Tongji Zhaiyao* 1994 [Summary of Chinese Statistics 1994] (Beijing: Zhongguo Tongji Chubanshe, May 1994), p. 30.

33. *Zhongguo Tongji Zhaiyao* 1993 [Statistical Yearbook of China 1993] (Beijing: Zhongguo Tongji Chubanshe, August 1993), p. 650.

34. *Zhongguo Tongji Zhaiyao* 1994 [Statistical Yearbook of China 1994] (Beijing: Zhongguo Tongji Chubanshe, November 1994), p. 530.

35. *Sichuan Jingji Nianjian* 1987, p. 257.

36. Ibid., p. 509.

37. *Guizhou Nianjian* 1987, p. 480.

38. *Guizhou Nianjian* 1989, pp. 396–397. In 1992, the Zhenhua Electronics Conglomerate was named as one of fifty-five large enterprise conglomerates and given a separate listing in the economic plan. *Guizhou Nianjian* 1992, p. 335.

39. The author would like to thank Ricky Tung, a colleague at the Institute of International Relations, Taipei, for pointing this out. Drainage of funds is part of the drain on state assets that has become a cause for concern to the Beijing authorities in recent years.

40. *Guizhou Nianjian* 1993, p. 354.

8

A Business Perspective

Richard J. Latham

When we talk about China's defense industries and defense conversion, are we studying a military or an industrial subject? To date, most Western studies have taken the military or political point of view; there have been few Western analytical contributions from the business perspective. Similarly, Chinese commentators, like their foreign counterparts, are largely defense intellectuals.[1]

The question is not of the exclusion of industrial executives and managers but rather of one's craft. The scholar or analyst must publish or perish, but the industrial manager must manufacture or perish—and thus has little time (or patience) for well-turned phrases or closely documented analysis.

China's defense industries, like all Chinese state-owned enterprises (SOEs), are now intensely preoccupied with the manufacture or perish challenge. Defense industries face a tricky challenge: They must manufacture less of one thing (military goods) and more of another (products suitable for the civilian market). Because of this economic imperative, with its complex and sometimes contradictory outcomes, we come to the inescapable conclusion that the study of China's defense industries principally involves commercial, economic, and industrial issues.[2] We have placed too much emphasis over the last three decades on the military aspects of these industries. If we do not take into consideration the watershed changes occurring in these industries, we risk failing to grasp the key dynamics transforming them.

New Organizations

From the late 1950s until about 1979 it was possible, if incorrect, to view China's defense industries as an extension of the People's Liberation Army (PLA). Indeed, there was much confusion about the nature of the industry itself, in part due to imprecision in Chinese terminology.[3] Nevertheless, it is essential to distinguish between PLA-run industries, which service the immediate logistical needs of the army and are known as "military goods enterprises" (*junxu qiye*), "army-run enterprises" (*junban qiye, jundui qiye,* or *jungong qiye*), or "army production enterprises" (*budui shengchan jingying*), and the much larger and important weapons-producing "ministerial defense economy" (*bumen junshi jingji*) or "ministerial defense industry" (MDI), which is composed of the ministries and, since the 1980s and 1990s, corporations under the civilian State Council.[4] Because these "corporations" are still more oriented toward policy than profit, they are referred to here as "ministries-cum-corporations." As one official explained in 1995: "Today the government wants us to behave like a corporation. Tomorrow they will expect us to act again as a ministry. Sometimes the expectations change several times in the same meeting."[5]

Beginning in late 1978, all aspects of China's socialist economy began to change as a result of experimentation with market forces. Eighteen years later, the forces at work within China's defense industries are stimulating or compelling far-reaching shifts in the ways defense industrial managers operate and their factories interact with the general Chinese economy. The primary actors are the newly formed corporations (*zonggongsi*) and conglomerates or groups (*jituan*) in the defense sector. The interests of these new organizations are increasingly linked to the national economy and to the internationalization of production. Their agendas diverge from those of central planners and sometimes conflict with the central government's objective of maintaining a core, state-owned defense industrial sector that concentrates on servicing military demands. They are driven by the twin imperatives of defense conversion and economic survival.

Defense Conversion: Military or Industrial Process?

Defense conversion—in Chinese, *junzhuanmin* (literally, military into civilian)—is an integral part of the transformation of China's defense industries. The PLA's Commission on Science, Technology, and Industry for National Defense (COSTIND) has been the institutional force behind the effort and has tried to involve international business. COSTIND has sponsored conversion conferences in Hong Kong and China in conjunction with the United Nations Development Programme, and COSTIND's leader, General Ding Henggao, and the U.S. Secretary of Defense William Perry have established a high-level commission that continues to address this subject. The interest of the United States was stimulated, in no small measure, by the large-scale downsizing of U.S. defense industries as a result of the end of the Cold War as well as the exceptional importance of maintaining communication channels and building confidence between the two powers.[6]

COSTIND's interest in defense conversion springs from several motives, including its oversight of and regulatory responsibilities for the defense industries. In addition, it may share some U.S. strategic objectives. The Chinese would like to counter regional impressions of China as an ambitious military power that possesses a massive arms industry, and defense conversion implies the overall diminution of defense-related activities. Moreover, there is no doubt that some Chinese policymakers and defense economists are intently focused on defense conversion. China does have a large and intentionally redundant defense industrial capacity; it arguably has far exceeded the requirements of the PLA since the mid-1960s. Thus, China in general is interested in conversion because the process means that the workforce remains employed. In this sense, defense conversion is probably more a public policy issue than a matter of national security interest or even macroeconomic policy.

Once we come to the industrial level, however, questions arise. Interestingly, there are minimal references to defense conversion in the trade newspapers of the former numbered defense industrial ministries. Indeed, segments of the PLA seem to talk more about the conversion of their army-run enterprises than do MDI leaders about their factories. There are virtually no discussions of or allusions to defense conversion among Western companies that undertake various forms of economic cooperation with China's defense industrial ministries. Finally, in neither the United States nor China have the government-to-government exchanges regarding defense conversion materially involved the executives or industrial enterprises that are most closely involved with conversion.

These ambiguities lead us to consider that former analyses of defense industries and defense conversion may have been misplaced and that there are differing definitions of conversion. Although the role of COSTIND is significant, COSTIND has more clout with army-run factories than it does with the MDI. What counts is not what happens in the bureaucratic superstructure or what the process is called but rather what happens in the industry. To understand conversion, we must first discover what the terms in which it is commonly discussed mean in the Chinese case.

Conversion, Restructuring, Downsizing, and Consolidation

Conversion

In the West, defense conversion involves the metaphorical transformation of swords into plowshares. This concept assumes that conversion results in fewer arms because there are fewer factories. Historically, however, Western factories have not been directly converted to civilian production. Rather, they have been shut down or consolidated, and employees find new jobs in other industries or professions. Market forces more than government policy have driven the process, despite official concern about the decline of the defense industrial base and the subsequent loss of defense manufacturing skill. Defense conversion—or, more properly, restructuring and consolidation—in the United States since 1987 has resulted in a 75 percent decline in the roster of defense suppliers, or a decrease from 120,000 to about 30,000 companies. More than 1 million jobs were eliminated. This figure is expected to rise by an additional 700,000 jobs by 1999.[7]

There has been no equivalent restructuring in China. Although PLA orders have declined and the subsidies required to keep the MDI afloat will also be cut, there is little evidence of a future reduction in the number of factories.[8] Market forces (i.e., falling PLA procurement) have precipitated manufacturing decisions within the factories regarding product mixes (for example, civilian versus military), but central government policies are not leading to structural conversion. In other words, the size of the Chinese defense industrial base has not been significantly converted, if conversion means fewer defense factories.

One can argue, however, that MDI factories have been substantially *diverted* to civilian production even as they keep defense production as their raison d'être. MDI managers regard civilian production as a cash cow that helps underwrite largely loss-making defense manufacturing programs. In part, the lack of permanent structural change is a reflection of the fact that China's defense industries are still defined by national boundaries. Western defense industries and governments have had to come to terms with the internationalization of arms manufacturing, whereas the self-reliance model—plus the tradition of state intervention and direction—still dominates Chinese thinking and constrains change. It is not clear that MDI factory managers, ministries-cum-corporations, or PLA policymakers are reconciled to the belief that "smaller is better." On the contrary, the "big and complete" mentality remains strongly embedded.

Nevertheless, the future is likely to bring true structural changes in the size and organization of China's defense industrial base. MDI officials have said that Beijing will approve the formation of *jituan,* which will be composed of civilian product lines that are hived off from existing MDI factories.[9] In other instances, defense manufacturing *jituan* may be formed on the basis of municipalities or provinces. In the near term, ownership may nominally be shared with the ministries-cum-corporations in Beijing, but most of the people in these corporations have not shown any obvious interest in making them truly entrepreneurial entities.[10] In the longer run (ten to twenty years), these newly emerging forms of ownership and industrial organization are likely to become increasingly unrelated to defense manufacturing. In effect, they will have been spun off. A core of MDI factories will remain, but that core will be much smaller than the current aggregate size of the MDI system.

Restructuring

Restructuring, more than production conversion, has been characteristic of recent trends throughout the U.S. and European economies. It certainly has been apparent in the defense industries in which restructuring was necessary because of lower demand for military goods and heightened competition for the remaining government business.[11] Restructuring is directly related to reducing production costs and is equally applicable to civilian and defense industries in which global competition has relentlessly driven down operating costs, prices, and profit margins. Restructuring typically leads to fewer jobs, simplified manufacturing processes, just-in-time manufacturing (e.g., lower inventory costs), the elimination of excess factory floor space, consolidation of manufacturing plants, and the formation of unified purchasing plans to reduce the cost of materials and services.

Restructuring as it is understood in the West has not been a common practice in Chinese conversion. Factories have not been closed, sold, or dismantled, and unused or surplus equipment has not been sold or auctioned. Capital assets, regardless of their age or utility, are typically retained—even unused, obsolete equipment and facilities. Because of the lack of competition and because labor is accounted for as a fixed cost, manufacturing processes have remained largely unchanged.[12] With an existing surplus of factory workers and staff, few economic advantages can be realized from streamlined manufacturing processes. There also has been little evidence of consolidating redundant manufacturing done at different factory sites. On the contrary, current examples of further subdivision of the existing defense manufacturing sector can be found. Unified purchasing already occurs, in part, within the ministries-cum-corporations as a result of quotas for raw materials. For above-quota purchases, each MDI factory obtains its own additional raw materials and supplies through regional markets, connections, or the bartering of its own surplus of allocated materials. Consolidated purchasing by the *jituan* and *zonggongsi* is a distant goal because these entities are still coming to terms with market economics.

In short, restructuring as a means of reducing costs and improving competitive advantage has not been a vital dimension of the Chinese defense conversion process. MDI managers understand the need for cost cutting; nevertheless, as workers for the central government they are hamstrung by numerous and interrelated fixed social welfare costs. Moreover, there are few variable costs that can be adjusted.

Downsizing and Rightsizing

As a Western practice, downsizing has a pejorative connotation. Often Western corporations have reduced personnel and reduced capital assets only to discover later that employees had to be rehired or capital assets needed to be repurchased in a seemingly endless cycle. The concept of rightsizing, however, implies rational, commercial decisions defining an acceptable, smaller size for a defense factory or corporation. Critics of downsizing suggest that rightsizing involves a more carefully thought out process of aligning cost, processes, management structure, and market demand.

Little publicized evidence exists of downsizing or rightsizing among China's MDI factories. China's defense industrial base is overcapitalized, overstaffed, and underutilized. Despite its shift to civilian product manufacturing, the MDI system remains largely intact. In the case of China, there has been no explicit public decision regarding an acceptable right size for its defense industrial base. Each factory, institute, and ministry-cum-corporation seems intent on maintaining all of its capital assets. One Chinese defense industry official said: "They [i.e., factories, institutes] are all our children. We cannot abandon any of them."[13]

Maintaining the Defense Industrial Base

Some writers in the Chinese journal *Junshi Jingjixue (Military Economics Studies)* seem to imply that when the civilian product output value of MDI factories exceeds 70–75 percent, central government planners should become concerned about erosion of the defense industrial base. Erosion includes actions of MDI factories ranging from ignoring defense production obligations to being unable to meet those obligations because the factories attached a higher priority to civilian manufacturing. It is unclear at what level this implied threshold of concern is measured: defense industries in the aggregate, *zonggongsi, jituan,* product sectors (for example, airframes, tanks), or individual factories. For example, the armaments industry (e.g., munitions) was called to task in the early 1990s for being too preoccupied with civilian manufacturing to meet core defense manufacturing requirements. Thus, although Chinese officials may be unable or unwilling to explicitly characterize the right size of China's defense industrial base, they have at least a vague sense of when a core defense manufacturing capability is being threatened. Interestingly, no MDI system leaders will acknowledge that any conflicts between defense and civilian production priorities have occurred.

PLA Factories

Downsizing and rightsizing seem to be irrelevant processes for the military-run factories. Over the years these factories have grown incrementally and apparently without any centralized directives from the PLA's General Logistics Department. Because these factories were never intended to be core defense factories, it is impossible to rightsize this economic sector or any of its individual factories, and because their existence was never consciously planned, standards of right, wrong, or appropriate size do not matter. Downsizing is also irrelevant because it involves aligning cost with market demand and prices. The army-run factories have taken on economic and social lives of their own that exist beyond the socialist planning processes. In metaphorical terms, they are industrial illegal migrants: Despite the questionable means by which they entered the socialist economy, they have eventually acquired an economic presence and political reality that the PLA and Beijing cannot ignore.

Consolidation

In China, as in the West, industrial consolidation is one course of action companies may take in response to diminishing defense sales; however, there are important differences. In the United States, for example, antitrust laws complicate the process, particularly when there are only two or three competitors. China has no antitrust laws, but also there has been no competition among its defense manufacturers. Consolidations in the West are negotiated based on expected profitability. Consolidations in China, such as they are, have been undertaken to absorb loss-making enterprises. In Western economies, consolidations lead to layoffs, the sale of excess capital equipment, the reduction of redundant manufacturing processes and staff support functions, and the permanent closure of factories. In China's case, there is little evidence that redundant MDI factories have closed in any significant numbers, the manufacture of similar product lines is not yet consolidated in fewer factories, and research is not centralized in a smaller number of key institutes or academies. In some cases, however, Third Line factories are being relocated to major urban and coastal areas.

In 1995, Beijing began to actively promote mergers or consolidations throughout the SOE sector.[14] Profitable enterprises were encouraged to acquire the assets and the debts of loss-making enterprises. Nonetheless, no fire sales occurred. For SOEs, a merger involves little more than a bookkeeping transfer of assets—the state technically remains the ultimate owner. China's MDI factories pose a special problem: If most of them are essentially loss makers, what advantage is gained by consolidating them? In late 1995, one ministry-cum-corporation set in motion a plan to consolidate about a half-dozen successful civilian-product profit-and-loss centers that were located within larger, loss-making MDI factories. These profit centers would eventually become separate enterprises that, in turn, would be part of an independent *jituan.*

Parent defense factories resisted the initiative, however, because it left the core defense manufacturing business without viable sources of revenue. Its underlying rationale was not readily apparent. On the one hand, if the measure was implemented, it would result in downsizing (i.e., the parent factories would be smaller). Earnings would be consolidated within one economically healthy *jituan,* which would be expected to result in further investment diversification on a larger scale than any single profit center could have undertaken. On the other hand, core defense factories would be left with less than encouraging prospects of becoming sustainable and profitable businesses.

The Chinese understanding of consolidation is illustrated by a story told at the November 1995 UN- and COSTIND-sponsored defense conversion conference in Wuhan. A foreign business executive described restructuring and consolidation in his country, in which about 40 percent of the production capacity was eliminated. To ensure that the Chinese in attendance fully understood what was said, the interpreter added his observation that the foreign staff and workers were transferred to other production lines or factories. It was necessary to correct this interpretation as it seriously misrepresented to the Chinese audience the effects of restructuring outside China. After the point was clarified that there were no other defense factories or production lines to which employees could be transferred, a Chinese regional defense industrial official commented acerbically to the Chinese audience: "This is indicative of the capitalist countries. They show no concern for their workers." With no more than a momentary pause, the official then turned to the foreign participants and asked what they were going to do about establishing joint ventures and placing manufacturing work in China's low labor-cost defense factories.[15]

Defense Restructuring with Chinese Characteristics

Whether we are talking about defense conversion, restructuring, downsizing or rightsizing, consolidations or mergers, China's defense industries are in great flux. This is leading to entirely new concepts of ownership and corporate organizations and new patterns of commercial activity. An approach to manufacturing in general and defense manufacturing in particular that is different from those found in Western defense conglomerates, Japanese *keiretsu* networks, or South Korean *chaebols* is emerging.

As China continues its experimentation with a socialist market economy, the policies and initiatives implemented for its 107,000 state-owned enterprises will be applied to MDI factories. The general economic transformation of China has been so profound that the MDI system can no longer maintain its insular and protected existence, although entrenched interests in the factories and institutes argue against the changes that have taken place or are anticipated. Ten of the likely changes, trends, or emerging phenomena are discussed in the following sections.

Three caveats are in order. First, the changes are ad hoc and experimental. There are few indications that Beijing, COSTIND, or the MDI leaders have any kind of preestablished or overarching plan that guides defense conversion. Thus, the changes will occur differently in the various MDI sectors. Armaments factories, for example, may be among the slowest to change because munitions factories have a hard time finding marketable civilian products that they can profitably produce.

Second, the speed and breadth of change are already linked to the ability of the MDI factories to sell their traditional big-ticket products (for example, ships, satellite launch services, aircraft, vehicles) in the international markets. When there is international demand for these goods or services, the factories have a foundation on which to reinvest and diversify. The leadership style of MDI leaders is an additional factor in these sales. In some ministries-cum-corporations the leadership bodies are populated by officials whose business instincts seem stronger than their impulses to engage in centralized planning. These leaders have aggressively sought new opportunities, including foreign investment, and have moved decisively to close mergers. Profits from big-ticket sales have allowed some ministries-cum-corporations to undertake restructuring and consolidation. Some of the more forward-looking ministries-cum-corporations have used their profits to send staff members abroad to obtain MBAs as well as law and accounting degrees.[16]

Third, it will take no less than ten years and easily as many as twenty or thirty years to significantly restructure China's defense industries, assuming the political will to change continues to exist. Almost two decades have passed since China began its economic reforms. The MDI system was actively engaged from the outset; it has not changed as much as its commercial SOE counterparts, but the reform process is clearly still under way.

Jituan

The most economically significant structural change is the formation of business groups (*jituan*), a trend already well under way in China's larger commercial sectors. By becoming a *jituan,* a company or group of companies can engage in business investments well beyond its original business scope. By pooling the investment resources of partner companies, a *jituan* is able to enter into and pursue high-return investments on a scale that previously would have been impossible for any single partner. Thus, Aviation Industries of China (AVIC) sought in 1995 to form a *jituan* called the China National Civil Aviation Industries, which was to be composed of five profitable civilian product operations drawn from core MDI aviation factories. The China Aerospace Corporation (SAC) created the Sanjiang Aerospace Group (*jituan*) in the Wuhan area, principally to participate in the burgeoning automobile industry in that region. In some instances, *jituan* may issue publicly traded stock, and Western accounting firms have been used to prepare for initial public offerings.

Ultimately, the *jituan* will likely be a hybrid of hierarchical Stalinist economic ministries and traditional Chinese hub-and-wheel business organizations. As such, they will roughly be the Chinese counterpart of the Japanese *keiretsu* networks or the Korean *chaebol* networks (the latter are based on kinship linkages). In the case of *jituan,* the basis of networking will be the old ministerial connections. The Chinese *jituan,* in the context of the MDI system, will not be synonymous with the old ministries—they are at a subministry level. Different *jituan* not only may become rivals with each other but also are likely to become powerful competitors of the original ministries-cum-corporations if the latter decide or are allowed to take on entrepreneurial roles. There is no set formula for determining the composition of future *jituan;* brief discussions with Chinese officials reveal schemes based on cities, provinces, and even regions.[17] Other plans appear to be based on similar product lines. The process of forming a *jituan* is more similar to a merger than a consolidation.

The *jituan* model appeals to many Chinese mainly because it has the potential to achieve significant capital accumulation. Presumably, this more rapid accumulation of capital can be used by *jituan* for mergers with loss-making MDI enterprises or for entirely new factories that can hire employees from closed or bankrupt MDI factories. Nevertheless, it will be impossible to create large numbers of *jituan* within each MDI sector simply because the number of attractive or meaningful merger opportunities will be limited.

Perhaps the key underlying rationale for a *jituan* is a more rapid creation of wealth than can be currently achieved in more narrowly focused SOEs. In addition, the success of a newly established *jituan* would seem to depend on the degree of social capital and social trust in Chinese society.[18] In other words, will the constituent members of a *jituan* have sufficient trust among themselves to allow *jituan* executives to operate them as national or even international corporations? Can *jituan* leaders avoid the centripetal pressures of regionalism or the stovepipe parochialism that typifies their ministerial roots? The answers are not clear. Communist culture has significantly reduced the traditional Chinese family basis of forming companies (although one cannot overlook the network of *taizi,* the children of high officials). Executive appointments and factory manager appointments seem to be based on technical merit or political reliability. However, evidence remains of extensive regionalism as a powerful influence. An insularity much more pronounced than in some commercial *jituan* organizations continues in the MDI system.

At this time, the ministries-cum-corporations still look and act like government ministries. Almost nothing in their organizational structures, residual central planning responsibilities, or financial operations suggests that these organizations are intent on becoming capital-accumulating corporations. Consequently, the energizing force for the economic transformation of the MDI system clearly rests with the emerging *jituan*—as the *jituan* go, so goes the MDI system.

Planning

We know little about the Chinese defense planning process. Together, the published literature and conversations with MDI officials over the last ten years suggest a much less detailed, complex, and formal process than is found in many other countries. The ministries-cum-corporations essentially are planning bureaucracies for defense research, development, and manufacturing. These organizations have the same staff limitations that are imposed on ministries: They are allowed to hire between 450 and 500 employees. After routine staff support requirements (e.g., logistics, party and propaganda work, personnel, legal matters, training, finance, discipline inspection, retired cadre affairs, and foreign liaison) are met, there are relatively few staff positions left to cover the key tasks of program management and planning.

Assuming COSTIND is the intermediary between the military branches and the ministries-cum-corporations, two questions arise immediately: How much planning is necessary, and who does it? These questions are being discussed among Chinese defense planners and officials in the MDI system, but the answers are still not clear.

First, PLA service branches would like to have a more direct voice in articulating their military requirements. There are persistent rumors that COSTIND's planning role will be modified. In part, this has arisen because COSTIND has been dominated by people from the missile industry. Hence, the argument goes, defense development and procurement requirements in other areas have not been adequately understood or represented. Second, aside from the need for integrated defense development and procurement oversight—by COSTIND or a future oversight organization—the reasons for maintaining an extensive intermediate planning bureaucracy are unclear. Third, if individual MDI institutes and factories are responsible for the development and manufacture of weapon systems, it is also unclear why the PLA services or a PLA coordinating agency could not deal directly with these factories and institutes.

The rise of *jituan* as potentially powerful commercial enterprises significantly reduces the need for the "corporate" function of the ministries-cum-corporations. If involvement in commercial matters becomes increasingly superfluous for the ministries-cum-corporations, then only the planning function remains as a substantive activity. But the planning staffs and functions could be transferred to COSTIND or a new consolidated planning staff within the PLA. In short, the ministries-cum-corporations may not exist meaningfully beyond the next five to ten years. The economic performances of the *jituan* will figure importantly in the future of these planning vestiges of China's planned economy.

Redefinition of the MDI System

Whereas restructuring in most market economies is quite linear and direct, Chinese restructuring will be circuitous and less clearly evident. As the ministries-cum-corporations become more superfluous to the commercial economic activities carried out in the MDI factories, the historical system boundaries will begin to erode. The aviation sector, for example, will be defined by *jituan* organizations, product groupings (e.g., helicopters, transportation, fighters, etc.) that may establish internal profit-and-loss centers for civilian and military products, and increasingly independent subcontractors that now are ambiguously subordinate elements of the Airborne Equipment Systems Engineering Department of AVIC, the specialized aviation production enterprise group, and the airborne equipment manufacturing enterprise division.[19]

Planning for defense product development and manufacturing is the only function that currently binds the MDI factories to the ministries-cum-corporations. With more than two-thirds of the revenue-generating business of these corporations now derived from civilian products, however, the planning function has little effect on the majority of their economic activity. Because planning no longer matters much, the former defense industrial ministry system will be further pressured to face current economic realities.

In the future, ministries-cum-corporations like AVIC and China Aerospace Corporation will still have legitimate regulatory, planning, and oversight responsibilities. But just as airline operations can continue without direct management by the Civil Aviation Administration of China (CAAC, a non-MDI, ministry-level organization), MDI factories can operate as commercial enterprises without ministry-cum-corporation involvement.[20] Only habit and established mindsets stand in the way of this kind of change; there are no objective reasons the PLA and the defense factories cannot more directly carry out defense procurement and planning.

Growing Independence in the Ranks of MDI Factory Leaders

Over the past four years, MDI factory managers and staff have become less opaque in expressing objections to government interference in civilian commercial activities. Across a broad spectrum of the MDI system, there is ample evidence of this burgeoning independence. Whether this sense of local initiative is tacitly encouraged by the ministries-cum-corporations or initiated at the local levels (to the dismay of Beijing planners), however, is still unclear. In some ministries-cum-corporations, factories and institutes can deal directly with foreign companies in business matters related to civilian products. Many of these entities also receive and disburse foreign exchange, enter into joint ventures with foreign firms, and engage in manufacturing for local and export markets. Nevertheless, some ministries-cum-corporations had not yet received central government authorization to hold foreign exchange accounts several years after becoming corporations—further evidence that they are viewed as government ministries rather than as economic entities.

Nonconverted Civilian Products

A less well-defined area for entrepreneurial initiative involves "nonconverted civilian products," which include original equipment or components and whole systems that can be sold to civilians and military end-users. Original equipment manufacturers (OEMs) supply components such as motors, engines, wiring harnesses, radar, propellers, environmental control systems (ECS), brakes, air-conditioning, airline seats, and tires. Examples of whole systems are jeeps, trucks, small watercraft and ships, small transport and passenger aircraft, communication equipment, and avionics. These items or systems are labeled "nonconverted civilian products" rather than "dual-use" products because the latter term implies a more sophisticated level of technology and a degree of controversy when foreign technology transfers are involved. In the case of the former kinds of products, the MDI factories do not initiate entirely new product lines. Rather, the factories use existing production lines to respond to civilian commercial demands.

During the mid-1990s, the MDI planning bureaucracies maintained an interest in this area for understandable reasons. There are potential conflicts between military procurement requirements and the factory-level pursuit of foreign exchange or higher domestic profits from nonmilitary customers. The central planners also have a responsibility to determine what new military products are developed. At the level of whole systems, the PLA was always the authorizing agency and the source of development funding. No factory possessed the independent financial resources to undertake anything not approved and funded by Beijing. It has become evident, however, that some MDI factories that made exceptional profits with civilian products now seek to use those profits to undertake selective investments in their core nonconverted product businesses, independently of their planning masters in Beijing. Because funding is supplied by neither the PLA nor the State Council, the central planners apparently cannot prevent or dampen these local initiatives to produce nonconverted whole systems.

In fact, the ministries-cum-corporations may implicitly welcome such activities. Success will breed success, but the implications for less aggressive or less financially successful MDI factories are great. Competition may force changes that the central leaders could not effect. The bolder and more successful factories will logically become the beneficiaries of military procurement. It is hard to imagine, for example, that planners in Beijing would compel the PLA to buy products from one factory when another produces a better product at a lower price.

Economic Rationalization of China's MDI System

If the MDI system is anything, it is organizationally and functionally rationalized. Unlike many of the army-run enterprises, which were established locally without much overall integration and planning (excluding the enterprises with four-digit designators), the MDI system was consciously planned. Unauthorized factories did not spring up as did army-run factories. There was a certain degree of redundancy intended: The system was designed to support regional hub-and-spoke manufacturing and not only reflected the strategic placement of factories to minimize the consequences of a ground war on Chinese territory but also formed a response to the weak infrastructure network of China in the 1950s. At the same time, due to the high degree of planning, there was extreme rationalization, with sometimes only one factory producing a particular input for the rest of the system. As a consequence, restructuring cannot substantially reduce the number of MDI factories without affecting core defense manufacturing. For example, China has only one propeller factory. Specific types of airborne equipment are not produced by hundreds of different supplier enterprises.

Rationalization based on economic factors rather than central planning documents will almost certainly result in consolidations and mergers. Although a specific policy has not been acknowledged publicly, the evident trend toward relocating Third Line factories and institutes implicitly recognizes that redundancy must be decreased and profitable manufacturing should be relocated in urban areas that are close to customer bases. The Sanjiang Space Corporation, for example, has relocated many of its people from remote space and missile production bases to cities such as Wuhan and Xiaogan. Their personnel have been reemployed in OEM factories and joint ventures to support the rapidly developing automotive industry in the area.

Because economic rationalization of the MDI system will be based on commercial imperatives, the process can take place only as quickly and extensively as other economic reforms are adopted. The formation of *jituan* will allow the consolidation of related production in a smaller number of factories, provided that the *jituan* are able to place workers in new or converted civilian manufacturing. Central planners are bound to play roles as protectors of the national defense industrial base.

Commercialization of Research and Development

The MDI research institutes are under great pressure from several sources. Central government funding for research and development (R&D) is being sharply reduced. In early 1996, Beijing announced that all nonadministrative, nonmanufacturing units under the central government would have their budgets cut by 10 percent. Other traditional sources of funding have already been lost. The MDI institutes have clearly moved toward a commercialization of their formerly restricted defense-related research services. For the first time they have begun to assess the cost of their work and therefore are attaching price tags to it. Commercialization has led to marketizing of their services, and they have begun to search for nonmilitary customers. The ministries-cum-corporations, COSTIND, and the PLA military branches continue to identify compulsory defense research projects and provide the funding, but defense research alone cannot sustain the institutes.

The MDI institutes are in a difficult position. Like the MDI factories, they are saddled with numerous fixed costs and defense obligations that they cannot ignore, yet their funding resources have been reduced. Many institutes are able to undertake civilian research that is commercially profitable. Others have formed small-scale manufacturing enterprises. As the institutes become more sophisticated in determining their actual research costs, the PLA will have to pay them substantially more. In turn, however, the PLA is likely to become more demanding about the value returned by the institutes. The MDI institutes have also sought to reduce fixed personnel costs by doing less recruiting or by encouraging excess staff to take civilian jobs. In one organization, this objective was achieved by limiting the salaries and bonuses of personnel the institute leaders wanted to encourage to leave. In another, the assignment of housing or housing upgrades was manipulated to encourage departures. In yet another case, workers whose spouses took financially more lucrative jobs in outside organizations were required to give up assigned housing.

The commercialization of research at MDI institutes has also given rise to divergent views about intellectual property rights (IPR). The fundamental question is who owns the IPR that results from work done in the institutes: the institutes or the ministries-cum-corporations. To date, the PLA apparently has not weighed in about its proprietary interests. The question of IPR ownership becomes even more complex as a result of the marketing of research. Do the consumers of MDI institute services own the IPR if they paid for the research? Increasingly, foreign companies are undertaking research in MDI institutes. If the foreign customer and the institute agree to share the IPR (e.g., the institute can use it within China; the foreign customer owns the rights in his country), how is the rest of the world market divided?

It is unclear whether the *jituan* phenomenon among factories and enterprises will also develop among institutes; there is no evidence to date that indicates institute *jituan* are being considered. A foreigner conversant in Chinese is impressed when traveling throughout China by the extraordinarily large number of research institute signs posted above the entrances to buildings and compounds. This large, subsidized research community probably exceeds practical commercial needs. Assuming there is resistance to consolidating institutes or even forming a *jituan* network, several other possibilities exist. Commercial *jituan* and factories may absorb some failing institutes. Some institutes may form subordinate companies based on successful civilian product research; personnel and overhead costs could be transferred to these companies. In still other instances, PLA service branches could take over institutes whose work is exclusively focused on military applications.

State Intervention

Despite the ascent of market and commercial values in China, Beijing remains committed to the fundamental philosophy that state intervention is or should be effective. As such, the MDI system will not totally disappear. A planning bureaucracy for defense research, development, and manufacturing will continue in some form, just as it has in the civilian sector.[21] MDI officials and PLA leaders have expressed concern about the "civilianization" of manufacturing in China's defense industries. In short, there is no evidence that the Chinese have lost all faith in the efficacy of planning and oversight. In regard to defense manufacturing, they are not so different from their counterparts in most countries.

International Arms Sales

None of the world's major military powers has been able to sustain its domestic arms industry without recourse to foreign military sales—and China will not be the exception. In fact, foreign military sales for China must play a far greater role than now may be apparent. If the trends or changes discussed above continue to their logical conclusions—and they should—China's core defense manufacturing business will increasingly stand alone as SOEs that exist only to serve PLA customers. Certainly, commercialization of MDI factories will cause the PLA to pay more for what it buys, but without large government subsidies the PLA cannot afford to underwrite the core defense businesses. Thus foreign sales are economically essential, and this will provoke complaints from other arms-exporting nations.

Exactly how the MDI will benefit from international arms sales is hard to quantify. International arms transfer data are aggregated on the basis of states. These data can provide evidence of international trends but not economic insights regarding specific companies. In the Chinese case, most PRC sales data are estimated by foreign analysts; furthermore, data for specific MDI companies do not exist. Although it is still impossible to determine the effect of military sales on the bottom lines of particular Chinese companies, obviously military sales represent a critical source of revenue. However, the positive impact of foreign military sales on the bottom lines of MDI factories may be somewhat attenuated by the reported direct foreign sales of "surplus" equipment by the PLA or other PLA companies. This practice will unnecessarily perpetuate the loss-making plight of MDI factories. Despite the arguments attributed to the PLA about its need to make money, the domestic economic reality is that Beijing has charged China's MDI factories with being economically self-sufficient. The MDI factories can neither serve the PLA well in the long run nor meet their profit objectives in the short run if the PLA competes with them for foreign sales.

The Internationalization of Defense Manufacturing

Offset and counter-trade obligations, more than any other factor, led to the internationalization of defense manufacturing. Until only a few decades ago, defense industrial bases in all economically advanced countries were contained within national boundaries. Defense manufacturers never willingly embraced offset obligations as a condition of foreign sales. Offset requirements have now resulted in substantial offshore manufacturing that is mistakenly attributed to the pursuit of lower labor rates. The oft-cited examples in the press of the location of parts manufacturing in China by Boeing, McDonnell Douglas, and various OEMs are direct results of offset obligations. Offset-based contracts have given Chinese manufacturers experience and new technologies. The benefits are illusory in the long run, however, because the Chinese factories are still vulnerable to contracts that require Chinese purchases of U.S. equipment. If there are no Chinese purchases, there are no offset credits.

Offset obligations have led to the diffusion of manufacturing and technology beyond the national borders of the high technology and major arms-manufacturing nations. Following on the heels of this dispersion of manufacturing—usually on a contract basis as opposed to direct foreign investment in manufacturing—have come manufacturing partnerships and alliances. The keys to this trend have been the high costs of development, the need to distribute development and sales risk, and the necessity that risk-sharing partners be able to deliver the markets of their respective countries. Principal defense systems, of course, are still routinely manufactured within national borders, but much of the nonconverted product manufacturing has been dispersed throughout the world. Although national self-reliance in defense manufacturing remains the ideal of many governments, the economic reality is that development, technology, and manufacturing take place internationally.

China has begun to wrestle with this trend. Although many military planners and MDI executives acknowledge these developments, strong impulses for Maoist self-reliance remain. In 1995, the United Technologies Corporation (UTC) was the first Western company to conclude an equity joint venture with an MDI factory for nonconverted manufacturing. The Chengdu Engine Company and UTC's Pratt and Whitney aircraft engine company formed a separate equity joint venture to manufacture aircraft engine parts. Boeing, McDonnell Douglas, and others contracted for parts to be made or assembled in China, but these arrangements have not yet involved joint ventures. AVIC announced that during 1996 it will make a decision regarding a risk-sharing partnership with a Korean company and an unannounced Western partner to develop and manufacture a 100-passenger aircraft called the Asia Express (AE 100). By early 1996, several European and U.S. companies had entered into negotiations for additional equity joint ventures with MDI factories to make nonconverted products.

Given the paucity of development funds in China, especially for low- or no-profit defense manufacturing, China's MDI factories cannot avoid establishing partnerships and equity joint ventures with foreign companies for the manufacture of nonconverted products. Nevertheless, the trend toward international partnerships and equity joint ventures creates difficult issues for China, especially because of recent expressions of extreme nationalism. International interdependence, particularly in matters related to defense manufacturing, runs counter to the Maoist tradition of self-reliance. In 1995 Chinese commercial nationalism showed a greater sensitivity to visible foreign brand names, foreign logos, and the perceived dominance of some product sectors by foreign equity partners. The sense of a foreign commercial invasion arose, not unlike the reaction in the United States during the major inflow of Japanese investments in the 1980s. Further, the pervasiveness of internationalized defense manufacturing, coupled with a Chinese sense of comparative technological disadvantage, is likely to intensify Chinese feelings of vulnerability or of being pressured and exploited.

No one can say exactly how the Chinese will work through these challenges, but in one form or another the Chinese must come to terms with the internationalization of nonconverted defense manufacturing. It clearly brings with it national and industrial vulnerabilities that Chinese and foreigners alike once eschewed and still fear. For the Chinese, the trend also brings direct access to international markets—it is a springboard to state-of-the-art technologies and partnerships that can lead to substantial dividends for the participating factories. Although political and national security policies will impinge on the decisions regarding changes in China's MDI system, at the grassroots level profits and losses of specific defense enterprises matter more. Unless the Chinese evolve a system of military and nonconverted product manufacturing that departs markedly from what exists elsewhere in the world, Chinese factory managers will find that their success is unavoidably linked to the international marketplace. Chinese enterprise managers and *jituan* executives who express this reality will become the new voices in Chinese domestic politics.

Foreign Direct Investment and Restructuring

As the internationalization of defense and nonconverted product manufacturing becomes an increasingly important factor in MDI restructuring, consideration should be given to the nature of foreign involvement. COSTIND-sponsored defense conversion conferences and MDI trade promotion literature clearly show that the Chinese are looking for foreign involvement, but the question is, what will attract the interest of foreign companies? The answers are relevant not only to any foreign involvement but also to the domestic commercial success of MDI factories.

Realistically, foreign companies may be able to participate in three areas: nonconverted products, civilian products, and product R&D. Products used only for defense are intentionally excluded here. For many reasons, the Chinese are unlikely to accept direct foreign participation in this product area. At the same time, it is highly unlikely, although not impossible, that any foreign governments would allow their corporations to transfer technology to such manufacturing joint ventures.

Foreign companies look at many specific factors as they reach decisions about investing in China. Foreign corporate representatives and joint venture managers routinely consider the following business issues when evaluating MDI factories (or, indeed, any Chinese SOE):

- Offshore manufacturing only versus equity investment
- Degrees of equity and control required and allowed
- Production for domestic or foreign markets
- Market access, distribution, and after-sales service capabilities
- Expertise and technology available at the MDI plants

How foreign companies evaluate these issues may differ substantially depending on which of the previously mentioned three product areas a company decides to explore. The responses also vary depending on whether foreign companies intend to be investors (i.e., equity partners) or simply enter into supplier-buyer relationships. Finally, product differences affect responses. What this matrix of products and issue areas leads to is this question: Why would a foreign company be interested in MDI rather than other commercial Chinese SOEs? There is nothing inherently attractive about defense industries that engage in defense conversion.

Nonconverted Products

The specific nature of an item is particularly important when considering nonconverted production at an MDI factory. A foreign company that never previously engaged in nonconverted manufacturing would probably not seek a relationship with an MDI factory to produce such equipment. The MDI factory would most likely lack the managerial talent to convert the technology into a salable civilian product. Conversely, an automotive brake manufacturer might consider a relationship with an MDI factory that makes truck brakes or brake pads for aircraft. In other words, how easy is the transition to civilian markets?

A foreign company must also decide whether to seek an offshore supplier-buyer relationship or establish a joint venture. To date, not more than one or two equity joint ventures have been established with MDI factories to manufacture nonconverted products. Most companies have simply sought offshore manufacturing, which is likely to involve limited technology transfer and investment. Indeed, in some cases, the MDI factory may actually pay for a license rather than simply contract to provide production services.

Equity joint ventures in nonconverted product manufacturing are sensitive issues for the Chinese because they effectively allow a foreign entity to participate in the MDI system. The natural follow-on issue is the permissible or desired level of equity participation by a foreign partner—and that issue, in turn, is tied to control of the joint venture. In the case of offshore manufacturing, the MDI factory normally retains full control (i.e., ownership) of the factory. In an equity joint venture, on the other hand, control of the separately organized joint venture must be negotiated.[22] The Chinese are no less sensitive to foreign "control" in their defense industrial system than are other countries. The foreign partners, of course, are equally concerned about protecting their investment.

Whether a factory's product is for export, the Chinese domestic market, or both is a key consideration for foreign companies. Obviously, a nonconverted product for export only should not pose a major equity control problem: It will have no impact on domestic defense manufacturing. If the product is intended for the internal Chinese market or a mix of export and domestic markets, then the matter becomes more complex. The internal market may be both civilian and military. The national security logic is fairly straightforward, but the underlying commercial grammar is not. Foreign companies are unlikely to make major investments in nonconverted manufacturing unless they want a strategic presence, partnership, or alliance with an MDI factory in the Chinese market. Accordingly, they will invest in a relationship that benefits both parties in the domestic as well as international marketplace.

The foreign partner will also presumably seek an early advantage over foreign competitors in the Chinese market in the form of market access. Joint ventures are the dominant form of cooperation because a manufacturing presence does not sufficiently guarantee success in China. Depending on the product, foreign investors may require associated distribution networks, established agent agreements, and after-sales service networks. In the case of civilian products such as automotive parts, room air-conditioning, or appliances, these could be important considerations. The MDI factories may lack the distribution and after-sales networks that their commercial SOE counterparts have long had.

Finally, the equity value of foreign expertise and technology comes into play. Although the Chinese vigorously seek equity contributions, they tend to undervalue technology and expertise when determining the worth of each investor's equity contribution. There is another complication as well: A higher level of technology and greater likelihood of military utility in a product will be more attractive to the Chinese but will make it more difficult for foreign companies to secure home government export approvals.

This problem may go beyond conventional government export controls. Export control offices worry about security issues, whereas companies are occupied with commercial issues. In the case of world class technologies that have no export controls, foreign partners assuredly will seek equity control of joint ventures. It is hard to imagine a company transferring such technology under license to a MDI factory simply to locate manufacturing outside the home country of the transferring company.

To summarize, foreign commercial interest in the nonconverted product manufacturing of MDI factories is most likely to come from foreign defense industries that also make nonconverted products. Offset obligations will drive the offshore manufacture of products exported from China. Strategic interest in the global defense industrial market and China's future nonmilitary civilian market will play an important role in leading to equity joint ventures. Foreign, nondefense manufacturing companies normally will be less interested in MDI factories for nonconverted product manufacturing—the entry costs and market barriers are likely to be too high.

Civilian Products

Foreign companies that seek a relationship with MDI factories to produce only civilian products have a simple and straightforward assessment to make: The most pressing question is why they would seek an MDI factory rather than a commercial SOE for civilian product manufacturing. In general, low labor costs are not the issue; labor costs are lower in commercial SOEs. Rather, MDI factories offer comparatively well-trained workers, a large number of technical personnel, highly capitalized factories, and, in some cases, highly specialized workforces.

The trade-offs between offshore manufacturing and equity joint ventures are less clear for civilian products. A foreign company with a specialized product, a narrow international market, and comparatively little capital to invest might find it more reasonable to transfer the technology than to start a joint venture. If the commercial relationship does not work out, both sides have lost relatively little. If a foreign investor does pursue a joint venture, the national security issue is of minimal importance. The foreign investors are most likely to be nondefense manufacturers.

Foreign companies must also decide whether to target the domestic or export markets, which brings up some unusual considerations. Offset obligation contracts normally do not cover civilian product investments. MDI factories have resisted giving credit for civilian product investments. The offset administrators also generally refuse to give offset credit for investments or purchases that are made in the factories of other ministries.[23] Because offset obligations are not a powerful force in this product area, foreign companies that seek specialized manufacturing competence for niche industries (e.g., medical equipment), unusual products (e.g., fiberglass racing car bodies), or high-quality, low-volume products (e.g., artificial joints, specialized casting, forging and machining) are likely to find value in commercial relationships with MDI factories. At present, such products are usually exported.

Overall, there are not a remarkable number of foreign investors that seek out MDI factories to manufacture civilian products for the Chinese domestic market. In fact, at several Chinese-sponsored defense conversion conferences, MDI and military-run factory managers expressed disappointment that there was little foreign interest regarding investment in their factories.

The big issues for foreign investors are market access, distribution networks, and after-sales service. In the case of China, the problem is not access to the Chinese market in general; rather, the question is whether or not the MDI factories have access to their own civilian marketplace. There are some notable and successful entries (e.g., the Changhe and Songhuajiang minivans that are manufactured, respectively, by the Changhe Aircraft Industries Corporation and the Harbin Aircraft Manufacturing Company) into the civilian commercial market.[24] In terms of many other products, however, MDI factories face stiff competition from established commercial SOEs that have more experience with marketing and understanding customer demands. The commercial SOEs usually have established distribution networks and agency agreements with retail outlets. Finally, although after-sales service is still an underdeveloped area of business in China, the commercial SOEs are in a better position to move quickly to seize market share. Therefore, foreign investors who wish to establish their brand names in the China market are more likely to look first at commercial SOEs.

The issue of management expertise and technology is also simpler for civilian products. Matters related to intellectual property rights, trademarks, and patents remain a concern, but the problems are no greater in MDI factories than elsewhere. If anything, the conservative and disciplined nature of leadership in MDI factories may lead to greater sensitivity to these matters—although this cannot be proved one way or another. Foreign management expertise is sought but is undervalued. As in the case of nonconverted product manufacturing, the major management challenge is making market-based manufacturing and accounting decisions part of the operating values of the joint venture, partnership, or relationship. Indeed, MDI factories are probably a management generation behind successful Chinese commercial SOEs. Their managers must come to terms not only with the socialist market economy and foreign practices but also with the Chinese civilian market itself.

Product R&D and Testing

MDI research institutes and universities offer unique opportunities for foreign companies. Some foreign companies have established small joint ventures or manufacturing agreements with MDI institutes; however, research collaboration is the principal form of relationship. China's MDI research institutes are one of the pleasant contradictions one encounters in China. On the one hand, they are more closely linked to the Chinese military than are the factories. Ostensibly, the MDI institute people would be more conservative, security conscious, and less open to encounters with foreign companies. On the other hand, the people in the MDI research institutes are among the most cosmopolitan, well-traveled, informed, and educated people in the MDI system. Many have earned graduate degrees abroad. Thus, they are perhaps the most forward-looking elements of the MDI system that deals with foreign companies.

Foreign companies have discovered the value of MDI research institutes. These institutes routinely have large technical staffs and costly but underutilized research equipment. European and U.S. companies have contracted with MDI institutes to do testing of metal and ceramic samples to determine if the results from their own laboratories are confirmed by independent testing. In this way, more product testing can be done without driving up development costs. Various foreign real estate and construction companies have contracted for wind tunnel testing of bridges, buildings, and television towers. Still other foreign investors have turned to MDI institutes to redesign and test components from their civilian products so they are more suitable for the Chinese market. The UTC Research Center contracted with the Beijing Aeronautical Manufacturing Technology Research Institute to develop industrial laser-cutting techniques for its manufacturing joint venture with the Chengdu Engine Company.

Military-Run Factories

This analysis has intentionally avoided addressing military-run factories.[25] They are not part of the large MDI system that makes up China's core defense industrial base; they primarily produce quartermaster goods and undertake maintenance and repair of military equipment. Apart from MDI factories that were transferred to the PLA in the 1950s and came to have four-digit numeric factory designators, much of the rest of the system of military-run enterprises grew by unsystematic accumulation. The majority of these factories lack the capital and human resources of the MDI factories. Additionally, they are even more removed from the domestic commercial networks for civilian products.[26] The military-run factories are thus far less attractive to foreign companies seeking equity investments than are MDI operations. These factories, however, may represent good opportunities for overseas contract manufacturing of civilian products.

Conclusion: Competition or Partnership?

China has undergone significant economic change in the last two decades. Most remarkably, the changes have been promoted and carried out in large part by one or two generations of political and industrial leaders whose entire professional careers evolved within a system of highly centralized planning. Regardless of the shortcomings of the recent economic reforms, the record of industrial change and progress is astonishing. The first generation of managers who began their working careers after the initial reforms in late 1978 are now entering the middle level of management. In short, the managers with the least ideological baggage are just coming into positions of real responsibility. From this perspective, the prospects for still more significant changes should be even greater than in the past twenty years.

China's MDI factories face a challenging future. The necessity of their involvement with foreign companies is probably greater than they are prepared to concede, even among themselves. Self-reliance and redundant capacity are luxuries they, like their Western counterparts, can ill afford to enjoy in the future. The Chinese will need to come to terms with complex and conflicting questions about equity participation by foreign companies in defense industries, IPR, joint risk-sharing partnerships, and the definition of a reasonable and affordable defense industrial base. Most of these questions involve business decisions, and their execution will be a function of other business decisions. Furthermore, the success of the resulting enterprises will also be a consequence of business decisions.

If only business decisions entered the equation, many of China's MDI factories could quickly strike some attractive deals—as well as some bad ones. In reality, however, the business decisions involving defense manufacturing and cooperation will be colored by issues of national secrecy, national prestige, domestic political and economic agendas, historical legacies, and the international political climate at any given time. These political factors will drive or constrain such cooperation in the capitals of foreign countries as well as in Beijing. In the end, more than a few decisions regarding foreign direct investment in Chinese MDI factories will be made for reasons quite unrelated to good or bad business sense. In this regard, China's situation resembles circumstances elsewhere.

China's MDI factories cannot compete domestically or internationally if the Maoist model of self-reliance overdetermines their future. As we enter the next century, homogenization of the global economy in general has already become evident in defense industries. There is every indication that the Chinese recognize and understand this process, but their factual understanding has not alleviated their apprehension about how best to blend domestic and international business, or how to work with foreign business partners.

As the world's defense industries continue to undergo substantial consolidation, the Chinese economy represents one of the last major growth areas for foreign investments. Nonetheless, China's defense industries face their own consolidation challenges. MDI factories and foreign investors both hold important cards: markets, partnerships, technology, capital, human resources. A few Western companies now recognize the latent industrial and technological potential of China's MDI factories and institutes, but even fewer foreign governments recognize the long-term implications for international business alliances. The Chinese and foreign governments alike are still more focused on domestic economics. The factories and investors that play their cards well today will have strategic relationships tomorrow, and in broader, historical terms, the cards each side now holds may not be the cards they hold decades from now. In the West, as in China, important commercial defense joint ventures will be approved or disapproved for political reasons that may have no basis in sound business sense, while other joint ventures will be concluded for the right commercial reasons. The resulting interdependence will lead to new levels of confidence building and sustained communication. Economic and technological advantages will be significant for the specific companies and for the relevant bilateral and even multilateral trade relationships.

Notes

1. For much of the last twenty-five years, analysts who followed Chinese defense industries were mainly Americans with professional roots in the government or security think tanks. A small second generation of analysts whose curiosity was stimulated as doctoral students now exists, but there are almost no new writers emerging from the ranks of government service. In a small way this change reflects the fact that China's defense industries now attract less military attention—and rightly so.

2. This chapter is based on empirical evidence and interviews and conversations with foreign businessmen as well as Chinese factory leaders and planning officials who oversee China's defense industries and institutes. The insights from foreign businessmen were obtained primarily from meetings sponsored by the American Chamber of Commerce in China, the monthly Aerospace Forum meetings,

the monthly Joint Venture Managers' roundtable discussions, and the Automotive Industry Joint Venture Managers' roundtable meetings. For obvious reasons, I have not identified specific people, companies, or ministries.

3. See Liu Huamian, chief editor, Junshi Jingjixue Cidian [A Dictionary of Military Economics] (Beijing: Zhongguo Jingji Chubanshe, December 1993).

4. These PLA-owned factories, mines, farms, and enterprises primarily provide employment for military dependents, supplement local unit budgets, provide "in-house" maintenance services for military equipment, and manufacture items like clothing and uniforms, utensils, and shoes. They lack the heavy machine equipment used by the MDI. PLA units use a four-digit designator (e.g., the 3501 PLA factory and clothing outlet store located a few blocks from Beijing's five-star China World Hotel). Factories and research institutes belonging to the State Council side of the defense sector use a three-digit designator and also have commercial names for the sake of carrying out business with foreign companies. The numeric designators for institutes are still used in Chinese trade publications, without reference to the "open" or commercial names. The continued use of numeric designators for institutes suggests that they have a closer relationship with the PLA than the factories. The following vignette illustrates the split-personality problem (e.g., defense and civilian customers) that can be found in the institutes. In one instance, a corporate spokesman in Beijing told a foreign delegation that they would visit a certain institute located in the interior of China. He referred to the institute by a commercial name that the delegation duly noted. When the delegation members arrived at their destination, they were met at the train station by officials who identified themselves as leaders from the "xxx institute." The foreign delegation leader said they were told in Beijing that they were supposed to see some other institute. The institute leaders replied: "Well, we are that institute. We don't use that name. In fact, we don't know anything about it. You can just call us the 'xxx institute.'" (The Chinese routinely substitute "x's" for numbers when they want to conceal the identity of units. I have taken the liberty of following that practice here.)

5. Chinese general managers still refer to the corporate staffers who accompany visiting foreign executives to their factories as "the government people from Beijing."

6. U.S. interest in this subject is puzzling, given the general lack of conversion success in the U.S. economy. Still, it is apparently felt that perhaps there are lessons to be learned from the Chinese experience. Furthermore, the U.S. Department of Defense has been the most forward-looking of all executive branch agencies in seeking a range of constructive engagements with the PRC. We also cannot ignore the personal relationship between Perry and Ding, which goes back to the late 1970s. Bruce Gilley, "Peace Dividend," *Far Eastern Economic Review,* 11 January 1996, pp. 14–16.

7. S. Chapman, "The Demise of the Industrial Base," *Air Force Magazine,* September 1995, p. 60.

8. The nature of the reduction in subsidies is unclear. Presumably, they will be phased in. Factory managers have taken Beijing's statements seriously. They expect Beijing to increase the pressure on SOEs to perform, but defense factory leaders with whom the author has spoken subtly suggest they do not think the central government has the political will to totally cut off their subsidies due to fears of labor unrest.

9. One foreign businessman who was close to some of the Chinese planners of a *jituan* said that the financial effects of the hiving process would be gradually introduced over five years. In this way, the parent companies would have time to adjust. Additionally, the new enterprises, which use factory space, housing, and support services provided by the parent companies, would be required to sign contracts to pay for those services and support. The separating process, therefore, is far from being clear and final.

10. The issue of ownership in MDI factories is hardly a clearly resolved issue. In MDI factories some equipment is visibly marked as belonging to the "ministry." Factory personnel have confirmed this fact and have added that other equipment is purchased directly by the factories and is therefore owned by them. The ministries-cum-corporations maintain inventories of major equipment for all factories and institutes. Although ministries-cum-corporations may technically own certain pieces of equipment, there is little evidence that they subsequently order equipment to be transferred from one factory to another even when there is an obvious need at another factory. This would seem to imply that they have only nominal ownership control. How ministries-cum-corporations, factories, and institutes would determine ownership of capital assets when *jituan,* joint ventures, or independent commercial entities are formed is also far from clear.

11. An important difference between the MDI factories and many Western defense industries is that the latter actively seek government contracts. The profit margins are not high, but they represent steady manufacturing business especially when there are limited alternative markets for the products developed at the request of governments. However, it appears that some MDI factories would probably prefer to get out of the business of supplying the PLA. Profit margins are low, and there are limited foreign military sales.

12. One of the roles of the ministries-cum-corporations appears to be "sharing the wealth" by distributing work, offset work, and foreign investment among the factories. Despite these efforts to redistribute resources, some factories have prospered while others have stagnated.

13. Interview with author.

14.. "Merger Trend Takes Shape," *South China Morning Post (China Business Review),* 11 January 1995, p. B1.

15. Workshop on New Business Opportunities in China: "Restructuring the Military Industry," Wuhan, November 7, 1995.

16. The attitudes of a ministry-cum-corporation toward the pursuit of advanced business skills and degrees for staff members is revealing of how the organization views its mission. If it perceives that it is still fundamentally a planning organization, then there is no reason to invest limited funds in business degrees—especially from foreign universities—that the organization will never be able to use.

17. One example might be China North Industries Corporation (NORINCO). Interestingly, as the Ninth Five-Year Plan gathers momentum, NORINCO has announced ambitious growth and internationalization plans to further develop its *jituan* network. In February 1996, the firm announced 20 percent annual growth goals, aiming to reach 80 billion yuan (about U.S.$9.6 billion) in sales by the end of the decade. Civilian production would reach 93 percent of total output, or 65 billion yuan (U.S.$7.8 billion); defense output would fall to 5 billion yuan (U.S.$602 million). Such specific numbers, even given in projections, are unusual. The company said its strategic goals include ranking among the world's top 500 companies and emerging "as an international company whose influence will be felt at home and abroad." Over the next five years, the company's plans include the establishment of four or five 10-billion-yuan (U.S.$1.2 billion) *jituan* in the automotive sector and fifteen new enterprises in auto parts and chemicals. See "China North Industries Group Strives for Business Expansion," *China News Service,* 6 February 1996.

18. Some of the more recent writing on the importance of trust in business is in Francis Fukuyama, *Trust: The Social Virtues and the Creation of Prosperity* (New York: Free Press, 1995); he argues that prosperity is a function of trust.

19. From the mid-1980s through the early 1990s, the airborne equipment factories were nominally under the control of an entity called the China Airborne Equipment Corporation. It was a corporation in name only and was synonymous with the former Airborne Equipment Bureau of the Ministry of Aviation. This is an

example of a ministry lacking the entrepreneurial expertise or authorized authority to spin off independent corporations.

20. What is happening in the MDI system has already taken place in an analogous manner in the Civil Aviation Administration of China (CAAC). For decades, CAAC was the regulator, planner, and operating arm of all civil aviation in China. Once it was clear that airlines could be operated independently of the regulatory and centralized planning functions, the regional CAAC branches became independently operating airlines. They now compete fiercely for routes, passengers, air cargo, and profits. They resist regulatory oversight and the remnants of centralized planning in such areas as decisions about aircraft purchases—although they cannot ignore oversight completely. Although the separation is still unfinished, Air China, China Southern, and China Eastern have all formed *jituan* organizations that allow them to engage in real estate development, hotel construction, tourism, taxi companies, and other kinds of business. With nearly forty airlines in China—most of which are unprofitable—alliances, consolidations, and mergers are becoming commonplace.

21. For instance, in the mid-1990s CAAC provided convincing arguments that it needed to regain more centralized regulatory control over airline safety, airport management, regional air traffic control, and aircraft procurement. CAAC officials argued that in the initial years of decentralized control, there were serious abuses, mistakes, and dangerous practices that arose as a result of the weakening of regulatory and oversight power.

22. When foreign companies enter a joint venture with an MDI factory, the foreign partner does not acquire an equity position in the existing MDI factory. A separate joint venture is formed. In principle, therefore, no foreign company can directly own part of a Chinese defense factory. There has been no interest in forming joint ventures directly with the ministries-cum-corporations. On the one hand, these disguised defense industrial ministries have no in-house manufacturing or business resources. On the other hand, these ministries have been sensitive to the signals that might be sent to competing foreign companies if they appeared to be too closely linked with any specific foreign companies.

23. In principle, offset obligations are a matter of national policy. In practice, however, administrators insist that purchases or investments be made in sectors related to the interests of the administrators. It is difficult, therefore, for foreign multinational corporations that make direct foreign investments in China to get credit for the total amount of their investments. For example, the Chinese administrator of offset obligations resulting from aviation sales to China will not give credit for investments made in the Chinese automotive industry.

24. The Changhe and Harbin factories did not form equity joint ventures. They buy unassembled kits from Japanese companies and assemble the vehicles in China.

25. There is a wealth of material on military factories in the Chinese journal *Military Economics Studies,* published by the Military Economics Research Center and the Institute of Military Economics. There is no comparable journal that is openly available that addresses the MDI system.

26. See "Practical Guide to Chinese Military Finance" editorial committee, *Zhongguo Junshi Caiwu Shiyong Daquan* [Practical Guide to Chinese Military Finance] (Beijing: Jiefangjun Chubanshe, 1993). The sections of this book dealing with army-run enterprises provide excellent insights regarding how army-run enterprises are expected to be managed.

PART 4

The PLA

9

The Chinese Army's New Marching Orders: Winning on the Economic Battlefield

Tai Ming Cheung

Since the end of the Cold War, armed forces around the world have had to find new mission responsibilities, ranging from peacekeeping to drug interdiction. Few, however, have followed the People's Liberation Army's (PLA) march into business. Over the last decade, PLA units have constructed a sprawling military-business complex. The military's headlong rush into civilian production has created numerous problems: Corruption, especially among senior officers, is growing; professional standards are falling; and localist sentiments and interests are on the rise. These negative developments are beginning to hamper the PLA's role as a military organization and as the guarantor of the Communist Party's hold on power.

This chapter examines the development, structure, and activities of the military-business empire and the impact of these business activities on the PLA's unity and operational readiness. Issues and questions that will be addressed include the following:

- *The evolution of the PLA's business complex.* Why did military chiefs decide to allow the army to go into business? How successful have military enterprises been in earning profits?
- *The administrative and corporate structure of the military-business complex.* How effectively can the central military authorities, especially the General Logistics Department (GLD), manage the commercial activities of military enterprises and units? What are the roles of the military regions and lower-level military organizations? How economically powerful are military enterprises becoming?
- *The range of commercial activities undertaken by military enterprises and units.* How wide-ranging are the business operations of the military-business complex? In which areas are they concentrated, and do they enjoy competitive advantages in these sectors?
- *The financial returns generated from the PLA's business activities.* How much profit do PLA enterprises make, and how are these earn-

ings divided among the firms themselves, their parent military units, and higher-level military authorities? Are these revenues sufficient to cover shortfalls from government appropriations? What proportion of military businesses are profitable and what proportion are loss makers?

- *The impact of the military's business activities on the PLA's unity and operational readiness.* How has the money-making mania affected the behavior and outlook of military personnel, especially senior commanders? How serious are corruption, individualism, and other related economic vices? Is the military's role as a fighting organization seriously affected by its involvement in business? Is there a growing disparity in living standards between units stationed in prosperous coastal regions and those in less developed inland areas? If so, could this lead to growing friction and rivalries among these units?

The Rise of the Military-Business Complex

The PLA's involvement in economic activities can be traced to its origins as a guerrilla army. Military units relied on growing their own food to meet basic subsistence needs.[1] Following the establishment of the People's Republic of China (PRC) in 1949, the PLA devoted less time to farming and more to training to become a professional army. Nevertheless, it was also required to assist in the country's economic construction, although these tasks were handled primarily by specially created engineering support units such as the Capital Construction Corps. Regular military units were left to divide their time between military training, political study, and sideline agricultural and production work.

The PLA's priorities began to change quickly after Deng Xiaoping took power in 1978 and launched an ambitious, costly program of economic development. Appropriations for other parts of the state budget were cut to pay for increased investment in economic projects, and the military was one of the biggest losers. The PLA saw its military budgets substantially slashed in real terms from the beginning of the 1980s.

PLA chiefs were ordered at around the same time to undertake a comprehensive review of their military posture—specifically, of whether the need to maintain a massive standing army still existed as the threat of a Soviet invasion began to recede. After lengthy deliberations, they came to the conclusion that China no longer faced imminent attack and could afford to reduce troop levels. At an enlarged session of the Central Military Commission (CMC) in 1985, it was announced that a million soldiers—or one-quarter of the PLA's total troop strength—would be demobilized.

These drastic cutbacks in defense appropriations and manpower levels led to an examination of the PLA's role in economic activities. With the huge reduction in troop numbers, a substantial portion of the PLA's logistical support and production capabilities was no longer needed. More important, the military authorities found it increasingly difficult to make ends meet with their reduced appropriations. As a result, military chiefs decided to allow military units to expand their participation in economic activities to acquire additional sources of income. PLA factories were encouraged to convert from military to civilian production, major military trading corporations were established, and previously banned areas of economic activity, such as the tertiary industry, were opened up to military enterprises.[2]

Military units rushed to set up companies: Military-affiliated enterprises and companies doubled in number from around ten thousand in 1985 to more than twenty thousand within a few years, and total turnover and profits from the military-business complex grew rapidly.[3] The number of military enterprises in the tertiary industry posted the most spectacular growth, increasing by a hundredfold between 1978 and 1992—from fifty to five thousand firms.[4]

This explosive growth in the PLA's involvement in business activities quickly led to the emergence of serious problems, including corruption, smuggling, speculation, and the illegal use of soldiers and military equipment for commercial purposes. The CMC was compelled to issue a set of regulations in 1989 detailing forbidden business undertakings. The "Ten Nos," as the regulations were called, outlawed army units' use of vehicles or resources for "speculation, profiteering and smuggling"; they were also not permitted to "lend, hire out and sell military vehicle licenses, bank accounts and blank invoices"; they could not "participate in panic buying of goods at higher prices, illegally buy up goods, profiteering and raising prices in any form"; and active service military personnel could not be employed by companies or engage in trade.[5] Additional regulations banning military personnel and units from speculating in the foreign exchange, stock, and real estate markets were issued in the past few years.

The GLD, which is directly responsible for overseeing the military's economic activities, has had to undertake regular rectification campaigns to purge unhealthy business practices and revamp the military-business complex. Crackdowns and audits of enterprises and units have been conducted almost annually since the late 1980s. These campaigns, however, have had little success in cleaning up the business complex. The GLD has been hampered by inadequate resources and the lack of an effective regulatory system to monitor the activities of military enterprises and units.

Deng's call for faster reforms during a trip to southern China at the beginning of 1992 sparked another period of high-speed, freewheeling growth for the military-business complex as well as the rest of the economy. As a result, corruption and other unhealthy activities became increasingly acute. This led to a decision by the CMC to launch a hard-hitting armywide rectification campaign in November 1993. The PLA's headquarters departments set up a leading group—headed by the director of the GLD—for this task, and ordered division-level and lower units to stop all commercial activities and close, merge, or transfer their enterprises to higher-level military units. In addition, the GLD overhauled its enterprise management system. Scores of work groups made up of senior general department officers were sent to ascertain that all military units and enterprises complied with the central directives.

Careful attention was paid to military enterprises located in coastal regions, particularly in special economic zones. Twenty-eight work groups from Beijing were dispatched to inspect enterprises in the Shenzhen special economic zone in the first half of 1994 alone. By October 1994, nearly two-thirds of all military enterprises in Shenzhen had been merged or closed, and half of the army-owned enterprises in Hainan Island and other special economic zones were similarly affected.[6] One reason for the specific crackdown on Shenzhen-based enterprises was to stem the growing tide of military companies flooding into Hong Kong.[7]

By the beginning of 1995, the rectification campaign had been declared a resounding success. The total number of military-run enterprises had been slashed by 40 percent, with nearly one thousand trading companies alone being forced to close. Strict regulations forbidding higher-level military headquarters units as well as division-level and lower combat units from running enterprises were drawn up and rigorously enforced. All active service personnel were said to have withdrawn from business activities and more resources devoted to sideline agricultural and production work.[8]

The Structure and Management of the Military-Business Complex

The military authorities have had a difficult time managing the business activities of military units and enterprises. Nondisclosure of commercial operations and underreporting of earnings were common practices among units, to avoid turning over as much as 80 percent of their income to higher-level logistics departments. A double-track management system has thus been established by the GLD to oversee the running of the military-business complex:

- *Administrative control system:* The Production and Management Department (PMD) under the GLD is responsible for administrative oversight of military enterprises. These departments extend down to the provincial military district and also to major operational units, such as group armies.

- *Enterprise control system:* Most large and medium-sized military enterprises must now be affiliated with a conglomerate that is owned by either logistics departments at the general headquarters or the military region command. Although the rationale for such integration is that it promotes economies of scale and horizontal linkages, it also allows tighter supervision.

Administrative Oversight: The Role of the General Logistics Department

Although the GLD has general responsibility for oversight of the military-business complex, day-to-day management is in the hands of the PMD.[9] This department is an army-level unit headed by a major general that was originally established to oversee the PLA's sideline agricultural and industrial production activities. The PMD's responsibilities include drawing up long-term PLA business development plans, inspecting enterprises, managing labor, and overseeing major production projects.

All military regions, service arms, and the Commission on Science, Technology, and Industry for National Defense (COSTIND) also have PMDs located within their logistics departments. They directly manage affiliated enterprises, factories, and other production units. Many of them have a factory and enterprise management bureau that supervises the running of military-industrial plants. Shenyang Military Region's PMD also set up an economic and trade bureau to oversee the barter and other foreign-trading activities of enterprises in its area.[10] Lower down the chain of command, group armies and—until recently—divisions also have production and management offices to supervise the business activities of the enterprises they own.

The PMDs have not, however, been able to keep up with the explosive growth of the military-business complex. Their administrative structures and management techniques are outdated, and they lack the manpower and resources to monitor the proliferation of enterprises. Reforms introduced in the early 1990s and energetically pursued in the 1993 rectification campaign have led party committees to assume a growing role in the management of enterprises.[11] In addition, clearer guidelines have been drawn up that define which units are allowed to and which are prohibited from taking part in business operations.

Other GLD departments also have responsibility for overseeing specific aspects of the business complex. The GLD's Finance Department supervises the financial accounts of military firms and has established a finance enterprise office to carry out this task. The Finance Department has also set up, in cooperation with the Industrial and Commercial Bank of China, special accounting centers in all military regions. These centers function essentially as banks, and all military units are required to deposit their earnings at these centers.[12] The Finance Department's audit teams regularly undertake inspections of enterprises and units to check their accounts.

The Military Supplies Production Department oversees the production of quartermaster supplies, such as uniforms, bedding, and office equipment, by the GLD's factories. Most of the enterprises, however, devote most of their attention to generating more profitable civilian goods, and military-related output accounts for a rapidly dwindling proportion of their total production.

Corporate Oversight: The Role of Military Conglomerates

Central and regional military authorities try to keep pace with the country's transition to a market economy by placing responsibility for the supervision of military-affiliated enterprises increasingly in the hands of holding companies and enterprise groups. These conglomerates, which are run as moneymaking ventures, are in many ways better suited to managing the military enterprises than are their GLD administrative counterparts. Most of these conglomerates have been formed in the past few years and are organized at three different levels:

- *National-level entities:* Each PLA general headquarters and arms service department has one or two conglomerates. Companies owned by units affiliated with these departments have had to turn over control to these corporations.
- *Regional- and provincial-level entities:* Each military region and some provincial military district commands have established conglomerates for enterprises in their areas of control.
- *Unit-level entities:* Group armies and units at equivalent ranks have also set up corporations to run their enterprises. These corporations vary in size and profitability, depending on their location and the operational responsibilities of their parent units. Those based in coastal regions tend to be more prosperous than their counterparts in inland areas.

National-level military conglomerates. National PLA conglomerates are among the largest and most profitable of all military-run enterprises. They were also among the first military corporations to be established in the mid-1980s. Each general and service headquarters department has one or more of these conglomerates:

- *General Staff Department (GSD):* The Poly Group Corporation and China Huitong Corporation are the two principal corporations that belong to the GSD. Poly was the first to be established (in 1984) and was originally known as Poly-Technologies. Set up as the foreign weapons trading arm of the GSD's Equipment Department, it operated publicly as a subsidiary of China International Trust and Investment Corporation (CITIC). In 1992, Poly emerged from the shadows to become an independent enterprise group with about one hundred subsidiaries under its control, including one or two listed companies on the Hong Kong stock market and trading firms in the United States and other countries. With assets of almost 8 billion yuan at the end of 1993, Poly has the largest capital base of all PLA enterprises and is also the PLA's top foreign trading entity.[13]

Huitong was established in the early 1990s and took control of all GSD enterprises apart from those that belonged to Poly. It has over six hundred factories, mines, and farms and more than fifty technological research institutes. It engages in a wide range of business activities, from running hotels and tourist companies to foreign trade. Although it has considerably more factories and workers than Poly, Huitong is much less profitable because a sizable number of its enterprises are loss-makers.14

- *General Political Department (GPD):* The GPD's major conglomerates are the Kaili Corporation and China Tiancheng Corporation. Kaili was set up in 1984 as a trading company for the GPD but has since expanded into many other lines of business. Its core activities include property development—especially in coastal provinces—and the production of audiovisual products such as laser discs and videos. Kaili has also been extremely active overseas; it has a listed iron-ore mining company in Australia, a bank in the Cook Islands, and business operations in Hong Kong.

  Originally established in the mid-1980s, Tiancheng became an enterprise group only in the early 1990s. It is now responsible for the oversight of all the GPD's enterprises, excluding those that belong to Kaili. It is estimated to have capital assets of around 4 billion yuan, has seventeen subsidiaries scattered around the country, and engages in a plethora of economic activities, from coal mining in Shanxi province to real estate investment in Guangdong province.[15]
- *General Logistics Department:* China Xinxing Corporation is the GLD's primary conglomerate. Xinxing includes nearly one hundred large and medium-sized enterprises employing 140,000 workers, as well as ten affiliated subsidiaries specializing in foreign trade, marketing, real estate development, chemicals, and advertising.[16] The GLD also has several other holding companies that manage the PLA's foreign trade relations. They include the San Ding Trading Corporation, which belongs to the GLD's Production and Management Department, and the GLD Trading Group.
- *Air force:* The air force's main commercial entity is the China

Lantian Industry Corporation, which oversees most of the air force's business operations. The air force has more than 430 factories and mines (although most of them are small), 160 air bases, and nearly 400 farms.[17] The air force also runs its own airline—known as United Airlines—which operates out of military air bases and uses air force transport aircraft flown by air force pilots.[18]

- *Navy:* China Songhai Industrial and Commercial Corporation is the navy's principal conglomerate and, like other army-run enterprise groups, engages in commercial activities ranging from industrial production to foreign trade. The navy also has its own shipping fleet, which uses naval vessels and operates out of naval ports.[19]
- *Second Artillery:* The Second Artillery's principal business venture is the Shanhaidan Enterprise Group, which is best known for its pharmaceutical products.[20] Another major enterprise group, China Tianlong Industrial and Commercial Corporation, may also be affiliated with the Second Artillery.
- *Commission for Science, Technology, and Industry for National Defense:* COSTIND runs a proliferation of corporations, including China Yuanwang Group, China Xiaofeng, China Galaxy, and China Xinshidai (New Era) Corporations. Xiaofeng was the first major corporation set up by COSTIND in 1984, although the company is said to have cut some of its ties with COSTIND in the past few years. Yuanwang appears to be COSTIND's principal corporation at present, whereas Xinshidai is a holding company that has kept a low profile in recent years, although some of its provincial subsidiaries have been more active.
- *People's Armed Police (PAP):* The PAP's chief conglomerate is the China Anhua Development Corporation, established several years ago as part of the PAP's Logistics Department. Anhua runs about seventy PAP enterprises and is authorized to engage in foreign trade. Before Anhua's establishment, the PAP conducted much of its business activities through China Jingan Equipment Import and Export Corporation, which is affiliated with the Ministry of Public Security.

These corporations are among the PLA's most profitable commercial enterprises. They receive special privileges from the government and military authorities, including wide-ranging decisionmaking autonomy, import-export trading rights, access to low-priced commodities, and generous tax and customs breaks.

Regional- and provincial-level military conglomerates. All the military regions and some provincial military districts have established their own comprehensive corporations in the past several years to provide commercial supervision for military enterprises and to take advantage of horizontal linkages to improve competitive advantages. For example, the trading subsidiaries of these conglomerates help to market the goods manufactured by industrial plants. Many of these plants are located in inaccessible inland regions and have been unable to find markets to sell their products on their own.

The conglomerates belonging to the military regions and districts include:

- *Huabei Jinghai Enterprise Corporation:* The holding company for the Beijing Military Region's enterprises, it has industrial, agricultural, mining, and other commercial operations in Beijing, Tianjin, Hebei and Shanxi provinces, and the Inner Mongolia autonomous region.
- *Dongbei Jincheng Industrial and Commercial Corporation:* Jincheng is one of the largest regional conglomerates and belongs to the Shenyang Military Region. It operates primarily in Liaoning, Jilin, and Heilongjiang provinces and has one thousand mostly small enterprises, a work force of eighty thousand, and registered capital of 2 billion yuan in 1992.[21] Another Shenyang Military Region conglomerate is the Songliao Enterprise Group, which specializes in vehicle production and repair.
- *Huadong Industrial Corporation:* Huadong oversees the enterprises belonging to the Nanjing Military Region. Huadong was set up in 1992 to improve coordination and integration among the military region's enterprises, which are scattered across five provinces and Shanghai.[22]
- *Shandong Dongyue Corporation:* Dongyue manages the commercial operations of enterprises in the Jinan Military Region, which covers Shandong and Henan provinces. There are more than six hundred army-owned enterprises in the military region, with fixed assets of 650 million yuan in 1994.[23]
- *Nanfang Industrial and Trading Corporation:* Nanfang is probably the largest of the military region conglomerates. Its enterprises are located in five provinces, including Guangdong, Hainan, and Hubei. Its most profitable and dynamic firms are in Guangdong, in particular in the Shenzhen, Zhuhai, and Shantou special economic zones and Hainan province. However, many of its plants located in inland provinces lose money. Nanfang was ranked as the country's 148th largest foreign trading corporation in 1993, with exports of U.S.$117 million.[24] The region's logistics department also oversees the Cuicun Enterprise United Group, which manages the military region's 500-odd industrial and agricultural production bases.[25]
- *Southwest Great Wall Economic Development General Corporation:* Southwest supervises the enterprises of the Chengdu Military

Region. With many factories located in inaccessible parts of the four provinces and autonomous regions that the military region covers—including Tibet, Sichuan, and Yunnan—Southwest is likely to be less profitable than its coastal counterparts.

- *Northwest Industrial and Trading Corporation:* Northwest runs the enterprises owned by the Lanzhou Military Region. Many of the military region's industrial plants are located in Shaanxi province, especially Xi'an, whereas farms and agricultural production bases are situated in more remote areas.
- *Provincial military districts:* A number of provincial military districts have also established conglomerates in the past few years to improve the management of their military enterprises. Sichuan Military District, for example, set up the Sichuan Bashu Enterprise Group in 1992 to assume the supervision of its commercial operations.[26]

Unit-level military corporations. The status of unit-level military enterprises is presently in flux because of the ongoing rectification campaign. One important decision made by the CMC at the beginning of the retrenchment in late 1993 was that all division-level and lower units would no longer be permitted to engage in commercial activities, although they could expand their sideline agricultural and production work. They were therefore ordered to close or transfer control of their enterprises to higher-level logistics departments. Although group armies have been allowed to continue to engage in commercial operations, they are also subject to stricter controls, which include turning over the management of some enterprises to superior logistics departments.

The rectification of enterprises in the group armies and lower-level units commenced toward the end of 1991 as problems over their freewheeling activities became increasingly apparent. A conference organized by the GLD in August 1991 discussed ways to tighten management over the business activities of enterprises and military units, including increased integration and encouragement of the development of enterprise groups.[27] Units at and below the division level were allowed to keep only two factories and were required to transfer control of their other enterprises to higher-level authorities. A hundred army-level and higher units were selected as "experimental centralization points" and were given management of these and other enterprises.

Not all group armies appear to have been included in this list of experimental centralization points. One unidentified group army in the Shenyang Military Region was reported in July 1992 to have turned only about one-third of its enterprises over to the centralization points.[28] This appears to have been the exception, however, as most other group armies have been greatly affected by this recentralization of control.

The nature of commercial activities among the group armies varies widely, depending on their location and operational status. So-called rapid deployment units, such as the 38th Group Army in Baoding, Hebei province, and the 39th Group Army near Dalian, Liaoning province, are likely to be less intensively involved in business activities because they are required to maintain higher levels of operational readiness than other units.

Group armies in coastal regions tend to do much better commercially than their inland counterparts because of the lucrative economic opportunities open to them, especially in the service and foreign-oriented industries. Their affiliated enterprises are able to engage in foreign trade, attract foreign investment, or undertake real estate projects. Military businesses located in inland regions concentrate more in industrial production, mining, and agricultural activities.

For example, the 42nd Group Army is one of the most successful of the PLA's armies. Based in Huizhou in Guangdong province, the unit runs a thriving conglomerate known as the Changcheng (Great Wall) Huihua Industrial Corporation, with more than ninety enterprises located in coastal open zones such as Shenzhen and Shantou.[29] Changcheng is heavily involved in foreign trade and cooperation with foreign investors. It set up a special industrial development zone near Shenzhen that has attracted several major foreign corporations.

The Luyan Enterprise Group, however, is typical of an inland-based, army-run corporation. Luyan belongs either to the 28th Group Army in Datong or the 63rd Group Army in Taiyuan—both in Shanxi province—and in 1992 had around ten enterprises, mostly factories and coal mines.[30] Luyan may no longer possess its mines, however, as the central military authorities ordered in 1993 that all military-run mines in Shanxi province had to be turned over to the provincial government.

Most of the enterprises owned by division- or lower-level units tend to be small ventures employing several dozens to several hundreds of people, mostly family dependents of military personnel or demobilized soldiers. Of the military's estimated twenty thousand enterprises, only a small fraction qualify as large or medium-sized ventures. Around sixty of the PLA's industrial plants are designated as large enterprises, each with assets of several hundred million yuan, out of a national total of thirty-five hundred large state-owned enterprises. A further 166 are officially categorized as medium-sized enterprises.[31]

Military Business Activities

Military-affiliated enterprises can be found in virtually every sector of the Chinese economy today. Before the onset of economic reforms at the end of the 1970s, almost all army-run enterprises were located in the agricultural and industrial sectors, but the focus has shifted rapidly in recent years, with military units flocking to more lucrative service industries. Official statistics suggest, however, that the economic focus of the military-business complex has not changed as much as it might appear:

- *Agriculture:* The output value of agricultural production by the PLA's 600-odd agricultural production bases and horse farms accounted for about 13 percent of total turnover from the PLA's economic activities in 1987.[32] The figure today is closer to 7–8 percent.
- *Industry:* Army-run factories, mines, and other industrial enterprises produced about 62 percent of the military's total economic output value in 1987.[33] Now they probably generate around 55–60 percent of the PLA's business turnover and around 40 percent of total profits.
- *Tertiary industry:* PLA enterprises engaged in the service industries—such as hotel management, medical care, trading and construction—produced 25 percent of the PLA's business turnover in 1987. This had officially risen to 30 percent by 1992, but given the enormous growth in the numbers of military enterprises involved in tertiary activities—in particular, many of the conglomerates—this may be a gross underestimate. A more realistic estimate may be that these enterprises account for 40–45 percent of the turnover and as much as 50 percent of the profits of the military-business complex.[34]

A heavy concentration of PLA enterprises can be found in the following industries:

- *Transportation:* An extensive transportation system dedicated to the military, including railways, roads, ports, and airports, has been commercially exploited by military units. Ground forces and navy transportation units have established large numbers of haulage and shipping companies.
- *Vehicle production:* Army-run factories produce around 20 percent of China's passenger cars annually. There are currently nearly seventy major plants that are capable of turning out fifty thousand vehicles and 500 million yuan worth of vehicle parts each year, although they manufactured only twenty-five thousand vehicles in 1993. Annual industrial output of all PLA vehicle factories in 1991 totaled 1.2 billion yuan, with posttax profits of 110 million yuan.[35] Design and production of many of these vehicles are substandard, however, and factories have been unable to sell them. Many of these factories will close or merge with larger units as China rationalizes this "pillar" industry.
- *Pharmaceutical production:* The PLA has almost four hundred phar-

maceutical factories, a quarter of which are large and medium-sized plants. They account for almost 10 percent of the country's total annual output of pharmaceutical goods, especially traditional Chinese medicines.[36]

- *Hotels:* Military units operate more than fifteen hundred hotels and guest houses employing more than seventy thousand workers. They range from spartan former military barracks to luxury hotels. Turnover from these hotels totaled 1.6 billion yuan in 1992.[37]
- *Property development:* Military units have joined the rush to develop real estate in recent years. PLA units occupy large tracts of prime land in many major cities, and many of them have teamed up with foreign and domestic property companies to redevelop these sites into high-rise office and commercial buildings. PLA units and their affiliated companies in Beijing, Shanghai, Guangzhou, Shenzhen, and Hainan have been particularly active in property development.
- *Textile production:* The GLD operates several dozen major factories that produce uniforms, garments, and other textile goods. PLA enterprises occupied seven of the top ten ranks in a 1992 survey of the country's largest textile enterprises.[38]
- *Mining:* Military units in fifteen provinces and autonomous regions run almost 150 major mines that produce coal and ferrous and nonferrous metals. Coal production in 1991 totaled 38 million tons, and a significant proportion was handed over to the government or sold on the market.[39] In 1993, PLA units were ordered to turn over all their mines in Shanxi province—one of the country's top mining regions—to the provincial authorities. This is estimated to have cut overall PLA coal production by one-third and resulted in lost earnings of as much as 1 billion yuan annually.[40]
- *Telecommunications:* PLA units have been able to use parts of the military's dedicated telecommunications systems for commercial ventures. PLA units have been especially active in the pager industry, and many military-run pager companies make use of military communications facilities. In Guangzhou, for example, three of the city's ten largest pager companies are owned by the air force, the provincial military district command, and a COSTIND subsidiary.[41]

The Earnings of the Military-Business Complex

The prime rationale for allowing PLA units to enter into business activities was to generate much-needed income to cover the military's severe budgetary shortfalls. The units appear to have done reasonably well. Official figures show that declared profits from the military-business complex are roughly equivalent to between 10 and 15 percent of the country's published defense budget. In 1992, for example, total profits from military business operations were 5 billion yuan (or U.S.$910 million at the 1992 official exchange rate[42]), and defense expenditures for that year were 37 billion yuan.

Income earned by PLA business ventures has increased rapidly since the mid-1980s. Annual turnover and profits have grown by around 15 percent on average for the past ten years, well above the forecasts of military authorities who had expected a more modest 5 percent annual increase in output value and earnings.

PLA enterprises located in the tertiary industry, such as hotels, trading firms, and property companies, have posted the most impressive economic results. In 1992, they accounted for 30 percent of total turnover from PLA business activities and 40 percent of overall profits.[43] The agricultural output of PLA farms has increased by a more modest 4–5 percent annually in the past few years.

The performance of military-run industrial enterprises has been the most disappointing. Around two-thirds of PLA factories are estimated to be running in the red, a higher figure than in the rest of the state sector.[44] Many of them suffer from outdated equipment, overstaffing, poor geographical location, mounting debts, and the inability to produce marketable goods. They were also hit hard by a nationwide austerity clampdown imposed by the government in July 1993 to cool the overheated economy. Bank loans were cut back severely, and many PLA plants found it nearly impossible to acquire working capital. Although this austerity campaign has been gradually relaxed, the government continues to keep a tight rein on credit supply, and many military enterprises still find it difficult to obtain bank loans.

Large enterprises, including conglomerates, have generally performed better than small and medium-sized firms. This is because they receive special privileges, including priority access to working capital, and enjoy economies of scale. The PLA's most profitable corporations are Poly, Shenzhen 999, and Xinxing. These three conglomerates may account for as much as one-third of the PLA's total business earnings.

The PLA's official figures for business turnover and profits almost certainly underrepresent what military-run enterprises actually earn from their commercial operations. Underreporting of profits is pervasive, especially among lower-level military units, and the military authorities have suffered serious losses in revenue collection as a result. The problem has become so acute that the GLD regularly conducts extensive audits of PLA-run companies and military units.

One self-audit conducted by all units at or above the regimental level in 1992, for example, found that a substantial number of soldiers had engaged in various "discipline-breaking problems, such as receiving more than their due, owning 'small treasuries,' arbitrarily paying living expense subsidies, bribery and bribe-taking." More than three hundred bank accounts containing illegally retained funds were discovered in the audit and confiscated by the GLD, although they were probably only a small proportion of the undeclared earnings hidden by military units.[45] The losses suffered by the military authorities in unreported business income could run into billions of yuan annually.

Military enterprises are allowed to keep a substantial proportion of their earnings. The amount varies according to the type of military enterprise and the profitability of their operations, but they are able to retain between 20 and 40 percent of their profits for reinvestment and other uses.[46] The remaining total is divided among the GLD (which receives the lion's share of between 40 and 60 percent), regional and provincial military authorities (which get 10 to 20 percent), and the local PLA unit that owns the enterprise (which gets 10 to 20 percent). If this division in earnings was applied to the 6 billion yuan that PLA units are said to have earned in profits in 1993, then the GLD received between 2.4 billion and 3.6 billion yuan.[47]

Enterprises pass on money to their parent military units for training and improving the living standards of troops. One group army in the Beijing Military Region said that it allocated 85 percent of the profits from its production activities to barracks construction and repair, food subsidies, construction of miscellaneous facilities, and medical coverage for troops.[48] But a substantial amount of the earnings is also used to entertain lavishly, purchase expensive luxury goods such as foreign sedans, and line the pockets of senior military officers and executives of the military enterprises.

The earnings generated by the military-business complex significantly help the military authorities survive inadequate defense outlays. According to PLA officials, about one-third of the current official defense budget goes toward living expenses, with the remainder allocated for maintenance and procurement.[49] Without the contributions from the military-business complex, the military authorities might have to devote as much as half of the official defense budget to meet living-expense costs.

Military-affiliated corporations also earn foreign exchange for the PLA. With a drastic drop in earnings from the weapons trade since the end of the 1980s, the military authorities have encouraged the rapid expansion of civilian exports. About fifty PLA corporations currently engage in foreign trade, and they exported U.S.$230 million worth of civilian goods in 1992.[50] Most of these earnings are retained by the GLD and probably finance the PLA's procurement of foreign arms, especially the expensive purchases of advanced weapons systems from Russia.

PLA enterprises may also be contributing a sizable amount of their output to help pay for Russian arms procurements. As much as half of the value of the contracts for the sale of Su-27 fighter planes, S-300 air defense missile systems, and Kilo-class submarines is to be paid in barter terms.[51] This could amount to several hundred million yuan annually.

The official PLA figures for earnings from the military-business complex are significantly lower than estimates made by Western analysts, which range from U.S.$5 to U.S.$20 billion.[52] Although the official PLA statistics for earnings are too low, it is unlikely that total profits made by the military-business complex go as high as even the most conservative of calculations by Western observers because many of the military enterprises run in the red or barely break even. Large and medium-sized enterprises are the most profitable, but they constitute only a small proportion of the total number of PLA enterprises.

A more realistic estimate for the PLA's annual commercial profits is about 10–15 billion yuan (U.S.$1.2–$1.8 billion) on a turnover of 50–75 billion yuan. The actual amount that may go to the GLD and other PLA units is probably only 4–6 billion yuan. Although the PLA's involvement in business activities has been lucrative, it may not be as much a cash cow as outside observers perceive.

The Impact on Military Unity and Center-Local Relations

The PLA's growing exposure to economic influences has encouraged a number of unhealthy practices that are beginning to erode the core values and institutional integrity of the military establishment. The most serious developments include rising corruption, falling levels of professionalism, the rise of "localist" sentiments among grassroots military units, and a widening wealth gap between coastal and inland units.

Corruption is perhaps the most serious and pressing problem. It is a price that the military authorities were willing to pay as long as the economic returns were satisfactory and the problem was limited to a tiny minority of soldiers. This appears to be the case so far, but as with the rest of the country, corruption within the ranks has spread rapidly. A *Liberation Army Daily* editorial pointed out in April 1993 that "at present, army-building is confronted with many new issues and challenges. A number of people in society are showing tendencies of making use of their power for private gains, mammonism, hedonism and caring only for profits. Such tendencies are also reflected in the army."[53]

The number of economic crimes prosecuted in the military judicial system since the late 1980s has risen rapidly. In particular, the growth in the number of prosecutions for serious offenses—those involving prison sentences of five or more years—is pronounced. This increase likely represents only the tip of the iceberg; many corruption cases never reach the courts and are dealt with internally by military units.

The authorities have expressed particular alarm at the growing number of senior officers who have been caught for corruption. In one of the most serious incidents in recent years, a deputy commander of the Jinan Military

Region helped to smuggle nearly one thousand luxury sedans from South Korea into Shandong province.[54] Another case involved the secretary-general of the Chengdu Military Region's Political Department, who was charged with corruption, speculation, and "dallying with women."[55] An anti-corruption drive shortly after the June 1989 Tiananmen Square crackdown is said to have brought in six corps-level and thirty-eight division-level officers.[56] In an investigation of its party members in 1988, the Chengdu Military Region discovered that although there were fewer general discipline violations than in 1987, more than half of the offenses involved corruption by party cadres.[57] Military chiefs had previously believed that younger soldiers, not seasoned senior officers, were most at risk from the "hostile international forces" who were trying to promote "peaceful evolution" in China.[58]

Although the senior officers caught for corruption represent only a tiny proportion of the total officer corps, the military authorities are worried that some of these "rotten apples" may hold key command positions. Because corruption is seen not only as an economic vice but also as an indicator of poor or weak character, it is thought that corruption will breed political unreliability. In the event of a crisis, military chiefs fear the actions of corrupt or politically unreliable officers could lead to a serious breakdown in the chain of command. During the 1989 antigovernment demonstrations, for example, the commander of the 38th Group Army, Major General Xu Dongxian, refused to order his troops into Beijing because he rejected the authority of the Party.[59] It is paradoxical that although business operations have become an essential lifeline for the PLA, they are also the backdoor through which ideas and influences that the Chinese leadership regard as spiritual pollutants enter into the military ranks.

In an effort to prevent the spread of corruption, the military authorities have ordered a clear separation between military units and the commercial enterprises that they run. These enterprises are not permitted to employ active service military personnel and may not be supported financially by military units. Before the November 1993 rectification campaign, these regulations were widely ignored and the military authorities had few means to effectively enforce them. The latest crackdown, however, appears to have curtailed the involvement of troops in business activities, albeit only temporarily.

Another major headache is the falling standard of professionalism among soldiers as a result of their focus on money making. The need to bring in income has forced some military units to devote less of their time to maintaining professional chores, in particular training. In any case, many units do not have sufficient funds to allow them to train regularly.[60] The lack of adequate training not only reduces the combat readiness and effectiveness of troops but also invariably leads to lax discipline and ebbing morale. In a conscript force with large numbers of poorly educated rural youngsters, military commanders already find it difficult to maintain adequate professional standards.

This problem of decreasing professionalism is made more acute by the peaceful security environment that China currently enjoys. With few prospects of the country becoming embroiled in any serious conflicts—especially involving the ground forces—in the foreseeable future, many soldiers will question the need to maintain a high level of military preparedness.

Because of China's improved security environment, however, a decrease in overall professional standards may not be as serious a concern for PLA planners today as it was during the Cold War. Changes in military thinking in recent years have shifted the focus of preparations from large-scale wars to limited regional conflicts. The PLA's modernization program now aims to develop a small but well-equipped number of rapid reaction forces that could be deployed to any part of the country within a short period of time. These elite forces consist of several group armies and additional division-sized forces. The rapid reaction forces get more funds from the central military authorities so they can concentrate primarily on maintaining a higher level of combat preparedness. The remaining frontline units, including an additional twenty-two group armies, will continue to be maintained at lower levels of readiness, which will allow them to pursue economic ventures.

At the same time as this two-tier operational force structure is taking root, the economic open-door policy is leading to growing disparities between military units across the country. Units stationed in fast-growing coastal regions and major cities do considerably better economically than their counterparts in inland areas. The central military authorities have taken some steps to limit these differences by providing additional subsidies to units in inland regions and by helping to relocate some army-run enterprises in remote areas to coastal regions or inland cities.[61] Nevertheless, these are token efforts; as the PLA's involvement in economic operations continues, these inequalities will become increasingly more significant.

As a consequence of this trend, the PLA's institutional identity may weaken as units become preoccupied with their own local economic priorities at the expense of central goals. Such a trend has already become apparent in the civilian arena, as prosperous provinces put their own interests ahead of those of the central government. Because the military is a more tightly managed institution, however, it will be more difficult for local sentiments to take root within military units. Although the high command in Beijing might seem far away for units stationed in Guangdong or Sichuan, its absence is filled by the military region commands.

As PLA units become more involved in business operations, local military-civil economic linkages have been growing more intimate. Local army-run enterprises are increasing their participation in joint ventures with civilian counterparts. These ties were once encouraged by Beijing; however, worried that this growing economic attachment might invite grassroots units to listen more to local leaders than to their military superiors in Beijing, the central military authorities moved to sever many of these local business ties. It was reported that military units were forced to cut ties with hundreds of local civilian enterprises during the last rectification campaign.[62]

Local authorities may also be required to take on an increased share of the military burden from the cash-strapped national government. Their contributions essentially consist of subsidies and the provision of other types of material support, primarily for military dependents and former soldiers. Local governments do not usually provide direct assistance to cover units' operational expenses or weapons purchases.

Warlord-style fiefdoms, however, are unlikely to arise. Many structural impediments exist to prevent local military commanders and civilian leaders from developing their own independent kingdoms. Military regions, for example, have no comparable civilian counterparts with which they would naturally cooperate. The province is the highest level of local government, which exercises wide jurisdiction compared to its PLA equivalent, the military district. Military districts wield limited authority, acting primarily as administrative organs responsible for militia and reserve work. They have little control over regular PLA units that are stationed in their province.

Localist tendencies are most likely to develop at the military subdistrict and county levels, involving militia, reserve, paramilitary, and other ad hoc private security forces. These organizations are not as tightly or carefully controlled by the military authorities, and their cores of full- and part-time military leaders are not believed to be reshuffled frequently. In addition, they receive limited funding from the central and provincial authorities and must find their own sources of financing or look to local authorities for assistance. There is no evidence, however, to suggest that military or local officials at this level have conspired to form their own independent power bases, although powerful feudal landlords have reemerged in recent years.

It should also be noted that only a relatively small proportion of officers and conscripts participate directly in business operations. Only personnel assigned to the logistics departments of military units are allowed to engage in such activities. Although these regulations have been widely ignored in the past and large numbers of troops have participated in economic activities, the military authorities claim that they halted this practice in their recent rectification campaign.

To prevent local military commanders from forging close ties with their civilian counterparts, they are now transferred at regular intervals to other parts of the country. Military region commanders and political commissars, for example, now serve on average two to three years before being moved. In the past, they often stayed in one post for twice this period of time.

Thus, despite the provinces' growing accumulation of wealth, the trend is not likely to threaten Beijing's control over the military. Although some local commanders and units might pledge their support to local leaders in return for economic rewards, most soldiers will remain loyal to Beijing. Even if the post-Deng central leadership was to become caught up in a divisive power struggle, the likelihood is small that the military would disintegrate and units would align themselves according to provincial or local interests. The PLA's command and control apparatus is highly centralized and well coordinated and can weather a crisis.

Future Trends

After a decade of rapid but often unhealthy growth, the Chinese military-business complex is undergoing a much-needed period of consolidation and restructuring. The military authorities want to root out serious structural problems before allowing further expansion. This rectification campaign began in late 1993 and was vigorously pursued throughout 1994. The crackdown appears to have been relaxed since the beginning of 1995, however, indicating that the retrenchment may soon be coming to an end.

The next stage of business development will differ from the indiscriminate nature of growth over the past ten years. With the rectification campaign and various other reforms in recent years, the authorities have signaled their intention to play a more active role in controlling business activities. Some features of this next phase of expansion are likely to include:

- *"Conglomeratization":* The authorities will focus on expanding the role of enterprise groups as the central institutions responsible for managing the activities of the military-business complex. More enterprise groups may be formed, especially at lower levels, and some of the present national-level entities may also be merged to form supersized conglomerates.
- *"Corporatization":* Another key reform will be to turn conglomerates and other military enterprises into limited liability corporations. This entails legally defining the rights and obligations of these entities, in particular their decisionmaking powers and the ownership of assets. Once their assets have been appraised and ownership rights clarified, they will be allowed to issue shares to their parent military units, higher-level military authorities, banks, other enterprises, and

any other parties that are owed debt by these enterprises. This reform is currently being pursued on an experimental basis in the state sector, and several military enterprises have been selected for this project. Although one of the goals of this reform in the civilian arena is to allow companies to publicly trade their shares, most military enterprises will be prohibited from doing so.

- *An emphasis on profitability and quality:* Only enterprises that are profitable will be allowed to continue to operate. Many smaller and older loss-making enterprises may be forced to close or merge with enterprises that perform better. This may affect thousands of entities. The surviving enterprises will be encouraged to concentrate their business activities in advanced areas such as high technology, vehicle production, and telecommunications.
- *Separation of functions:* Only units in the logistics system will be allowed to undertake business activities. Units in the combat and political control systems generally will not participate in economic activities, apart from sideline agricultural and production activities.

A thorough cleanup of the business activities of military units and enterprises will not be easy to achieve. The lure of making money is too tempting for many military units to resist. New bans preventing units from any future involvement in business activities may induce many of them to carry on covertly and may lead to the growth of an underground military-business complex that will be even more difficult for the military authorities to monitor and control.

Notes

1. China People's Liberation Army Historical Reference Materials Editorial Supervisory Board, *Houqin Gongzuo* [Logistics Work: Part 1] (Beijing: Jiefangjun Chubanshe, 1994).

2. For a comprehensive chronological account of the various high-level meetings and decisions on expanding the PLA's business activities during this period, see Jiang Siyi (chief editor), *Zhongguo Renmin Jiefangjun Dashidian* [Chinese People's Liberation Army's Major Events Directory] (Tianjin: Tianjin Renmin Chubanshe, 1992).

3. Former CMC Secretary-General Yang Baibing said the military had more than twenty thousand enterprises in 1992. See "Yang Baibing Speaks at Family Planning Meeting," Xinhua Domestic Service, 2 April 1992, in *Foreign Broadcast Information Service, Daily Report China* (hereafter *FBIS-CHI*), 8 April 1992, p. 41.

4. *Zhongguo Jingying Bao* [China Business Times], 29 July 1994, pp. 1–2.

5. "Army Units Banned from Commercial Activities," *Zhongguo Tongxun She* [China News Service], 4 April 1989, in *FBIS-CHI,* 2 May 1989, p. 116.

6. *Zhongguo Jingying Bao,* 19 August 1994, p. 7, and 21 October 1994, p. 7.

7. Interviews, Beijing, November 1994. Senior PLA officials have also told

visiting Hong Kong politicians that the National People's Congress was planning to ban soldiers at the future PLA garrison in the territory from participating in economic activities. See *Eastern Express,* 29 December 1994, p. 1.

8. *Jiefangjun Bao* [Liberation Army Daily], 10 February 1995, p. 1.

9. Ling Pingyi, He Guofa, and Wen Zifei, eds., *Jundui Shangchang Jingying Shiyong Zhengce Fagui Zongshu* [The Practice of Military Production and Business: A Summary of Policies and Regulations] (Beijing: Junshi Youwen Chubanshe, 1992), pp. 7–8.

10. *Jiefangjun Bao,* 15 May 1992, p. 3.

11. *Jiefangjun Bao,* 10 February 1995, p. 1.

12. "Army Establishes Accounting System," Xinhua English Service, 16 August 1989, in *FBIS-CHI,* 21 August 1989, p. 39; *Jiefangjun Bao,* 4 August 1994, p. 1, which reported on the fifth anniversary of these centers.

13. *Jiefangjun Bao,* 16 July 1994, p. 1.

14. Interview with a senior Huitong executive, Beijing, February 1994.

15. Interview with an executive from Tiancheng, Beijing, August 1994.

16. Company brochure.

17. China People's Liberation Army Historical References Series Editorial Oversight Board, *Kongjun* [Air Force], (Beijing: Jiefangjun Chubanshe, 1992), pp. 719–726.

18. Wu Yunhe, "United Airlines to Open Five New Routes and Buy 164-Seat Russian Aircraft," *China Daily* (*Business Weekly*), 28 February 1993.

19. Hu Chunhua, "Navy's Effort to Develop Maritime Shipping Service Has Broad Prospects," *Jiefangjun Bao,* 2 June 1992, in *FBIS-CHI,* 23 June 1992.

20. *Zhongguo Jingying Bao,* 20 May 1994, p. 7.

21. Company brochure.

22. Ding Chengren, "Huadong Industrial Group Corp Established," *Zhongguo Junzhuanmin Bao* [China Military Conversion News], 6 August 1992, p. 1.

23. *Zhongguo Jingying Bao,* 29 July 1994, p. 7.

24. Ibid.

25. *Jiefangjun Bao,* 17 January 1992, p. 1, and 3 May 1991, p. 1.

26. *Sichuan 1993 Nianjian* [Sichuan 1993 Yearbook] (Chengdu, Sichuan: Shehui Kexnue Chubanshe, 1993), p. 157.

27. *Jiefangjun Bao,* 4 August 1991, p. 1.

28. *Jiefangjun Bao,* 24 July 1992, p. 1.

29. *Jiefangjun Bao,* 1 December 1988 and 26 January 1992.

30. *Jiefangjun Bao,* 10 January 1992, p. 3.

31. *Jiefangjun Bao,* 20 September 1992, p. 1.

32. *Dangdai Zhongguo Junduide Houqin Gongzuo* [The Contemporary Chinese Army's Logistics Work] (Beijing: Zhongguo Shehui Kexueyuan Chubanshe, 1992), pp. 553–613.

33. Ibid.

34. The government has had great difficulty in accurately measuring the contribution of the service sector to the country's gross domestic product. In 1994, for example, it had to significantly revise the GDP growth rate after discovering that it had underestimated the growth of the tertiary sector by 20 percent. The service industry now accounts for around one-third of the GDP.

35. *Zhongguo Jingying Bao,* 28 October 1994, p. 7.

36. *Zhongguo Junzhuanmin Bao,* 8 January 1991, p. 1.

37. China Tertiary Industry Yearbook Editorial Board, *Zhongguo Desan Changye Nianjian, 1993* [China's Tertiary Industry Yearbook, 1993] (Beijing: Zhongguo Tongji Chubanshe, 1993), p. 879.

38. *China's Leading Companies, 1992* (Hong Kong: China Statistical Information and Consultancy Service Centre, 1992).

39. *Jiefangjun Bao,* 4 June 1992, p. 1.

40. Interviews, Beijing, May 1994.

41. Guangzhou Enterprise Evaluation Association, *Zhongguo Guangzhou Daxing Qiye Paixu* [Ranking of China Guangzhou's Large-Sized Enterprises] (Guangzhou: Zhongshan Daxue Chubanshe, 1994), pp. 115–117.

42. China Tertiary Industry Yearbook Editorial Board, *Zhongguo Desan Changye Nianjian,* p. 878.

43. Ibid.

44. Xing Zhengguo and Wang Shuanming, "Modern Enterprise Systems: Choice of Reform for Army-Owned Enterprises," *Junshi Jingji Yanjiu* [Military Economic Research] (September 1994), p. 82.

45. Wang Yihua, "Initial Achievement Made in All-Army Outlay Final Accounting," *People's Daily,* 31 May 1992, in *FBIS-CHI,* 11 June 1992, p. 26. In a 1995 audit, the GLD recovered nearly 1 billion yuan in undeclared earnings; see "All-Army Auditing Brings About 1 Billion Yuan in Economic Benefits," Xinhua Beijing, 16 December 1995.

46. Interviews, Beijing, February and April 1994.

47. Interview with foreign military attaché, Beijing, March 1995.

48. Chen Hui, Li Huamin, and Luo Tongsong, "Doing Credit to Army of Republic: Notes on Party Committee of a Certain Group Army Under the Beijing Military Region," *People's Daily,* 30 July 1991, in *FBIS-CHI,* 8 August 1991, p. 16.

49. Interview with foreign military attaché, Beijing, March 1995.

50. China Tertiary Industry Yearbook Editorial Board, *Zhongguo Desan Changye Nianjian,* p. 878. This figure indicates that the trade dealings of the GSD's Poly Corp. are not included.

51. Russian military officials say that barter goods they receive from the Chinese must be of sufficiently high quality that they can be resold for hard currency. The financial terms for the first batch of Su-27s that China purchased in 1991 were 35 percent hard currency and 65 percent barter goods. The spiraling cost of these barter goods within China in the past couple of years forced Beijing to renegotiate the barter terms of future contracts with the Russians in 1994. Under the new arrangements, the Chinese will pay between 50 and 90 percent of arms contracts in hard currency, the exact proportion to be determined by the size of the contracts and whether they are for weapons systems or technology transfers. Interviews, Beijing, September 1994.

52. This range of estimates can be found in David Shambaugh, "Wealth in Search of Power: The Chinese Military Budget and Revenue Base," paper presented to the conference on Chinese Economic Reform and Defence Policy, July 1994, Hong Kong; and Richard A. Bitzinger, "China's Defence Budget," *International Defence Review,* February 1995, p. 36.

53. "The Army Must Take a Good Lead in Building Spiritual Civilisation," *Jiefangjun Bao,* 16 April 1993.

54. Interviews, Beijing, December 1993. He was subsequently retired and apparently committed suicide shortly afterward.

55. Ji Wen, "Ning Yinghai, Former Secretary-General of the Chengdu Military Region Political Department, Sentenced to Imprisonment for Corruption, Speculation and Dallying with Women," *Liberation Army Daily,* 7 October 1988, in *FBIS-CHI,* 24 October 1988, pp. 32–33.

56. *Hong Kong Standard,* 3 April 1990.

57. "Central Military Commission on Corruption," Xinhua Domestic Service, 2 April 1988, in *FBIS-CHI,* 4 April 1988, p. 59.

58. See Lin Folian, "Beware of Soft Sell from Outside," *Jiefangjun Bao,* 2 November 1989. The writer points out that the external reactionary forces "use decadent capitalist ideologies and thinking on politics, economics, social ethics and indi-

vidual values to subvert and pollute the minds of our people, particularly the young ones."

59. Interviews, Beijing, July 1989.

60. See *Jiefangjun Bao,* 16 March 1990. The article points out that units use the excuse that there is a shortage of funds to cut their training programs.

61. See Guo Chunsheng and Ma Chunlin, "PLA Logistics Department Director Zhao Nanqi on Army's Production and Operations in New Fiscal Year," *Liberation Army Daily,* 19 January 1992, in *FBIS-CHI,* 10 February 1992, p. 29.

62. *Jiefangjun Bao,* 10 February 1995, p. 1.

10

Market Competition and the Management Systems of PLA Companies

Min Chen

There is an old popular Chinese saying that reflects the Chinese view of business competition: *Shang-chang ru zhan-chang* (the marketplace is a battlefield). Although the players in this commercial battlefield are not literally armed for battle, the intensity of their competition simulates a fight for survival. Since 1978, such competition has intensified as China's armed forces, the People's Liberation Army (PLA), are fighting on a "new battlefield." A widely known joke in China declares: "China has now only a navy, because the air force and the army have all 'jumped into the sea'" (*xia hai,* gone into business). Inexperienced and ill-prepared as they are, many PLA corporations have proven to be among the fiercest competitors and have recorded the most rapid growth.

Indeed, the PLA is no doubt unmatched among world armies in terms of the extent of its direct participation in business activities. Because of the opportunities resulting from the tidal wave of Deng Xiaoping's economic reforms, the PLA has been able to engage in a wide range of new activities, including weapons transactions, large-scale manufacturing, transportation, hotel management, import and export, wholesale and retail business, real estate, futures and foreign exchange dealings, and even illegal smuggling. One may argue that the PLA is currently the most business-oriented army in the world and financially the most self-reliant. In this chapter I will thus focus on the following four major topics: the management system of PLA companies, their competitive advantages and disadvantages, the formation of PLA conglomerates, and the internationalization of PLA companies. In conclusion, developing trends of the PLA's business will be briefly analyzed.[1] First, however, I must review the Chinese army's experience in self-support and commercial dealings.

Background

By any criteria, the PLA is not entirely a novice in market competition. The participation of its military units in self-supporting economic activity has long been the tradition of the Chinese Communist Party, starting with the Jiangxi Soviet and, later, Yan'an in the 1920s and 1930s. Despite the anti-Japanese front forged after 1936, the Red Army was poorly supported, leading Mao to point out in 1939 that "there are now only two roads to follow: one is to disband the army; the other is to be self-reliant." Mao concluded that in his view: "The second road . . . is the only way out. We should be engaged in production and save ourselves by being self-reliant."[2]

Thus in 1939 the Communist army began the famous "Big Production Campaign." Senior Communist leaders such as Zhou Enlai (Chinese premier from 1949 to 1975) and Zhu De (commander-in-chief of the Communist army until his death in 1975) took the lead in participating in agricultural production and textile weaving together with rank-and-file soldiers. The 359th Brigade led by the former PRC Vice President Wang Zhen, a role model at the time, allegedly went so far as to plant opium for profits to compensate for an inadequate military budget. The Big Production Campaign was deemed a great success in overcoming the blockade imposed by the Nationalist government during the anti-Japanese war.[3]

After the Big Production Campaign, the Communist army largely maintained the tradition of involvement in business whenever the circumstances required it. Since the founding of the People's Republic of China (PRC) in 1949, the Chinese government has invested heavily in defense industries in terms of capital, technology, and equipment; the responsibility for management of these industries, however, lies mainly with the State Council, which dominates both defense-related and civilian industries. Although the PLA maintained a close relationship with most of these industries, it participated in the direct management of only a few. The business activities managed directly by the army were largely restricted to agricultural and herding farms. Large-scale participation by the PLA in business has been a phenomenon of only the past decade.

The 1980s: Expansion and Opportunity

In the 1980s, China's economic reform and modernization programs were in full swing, and the whole country had caught "business fever." According to a popular saying: "Nine hundred million out of 1 billion people are engaged in business, with only 100 million standing by and watching." Modernization of the military moved at a snail's pace, due to a lack of funding, whereas modernization of the civilian sector progressed rapidly.

Deng Xiaoping addressed the problems of military modernization in a speech to the Central Military Commission (CMC) in 1984 on army participation in business. According to Deng, "Army construction should follow the overall constructive plan of the country." He called for the army, air force, and navy as well as the Commission on Science, Technology, and Industry for National Defense (COSTIND) to devote a significant portion of their resources to the support of national economic development and construction. The CMC issued the directive of "releasing water to feed fish," a green light for the armed forces to participate in business in order to compensate for deficiencies in the national defense budget.[4]

Thereafter, many PLA units established companies in various sectors of business and special production and service departments. Their business activities are found not only throughout China but also in Hong Kong, Macao, and overseas. By the mid-1990s, some PLA companies had already developed into multinational conglomerates, whereas many others had become leading business groups in certain areas of the country. Like millions of other business entities in China, PLA companies enter into almost any venture in which there is money to be made. Currently, PLA companies are involved in large-scale infrastructure construction, the management of luxury hotels, satellite launching, international weapon sales, amusement park development, telecommunications network operations, and transportation services. According to the *Economist,* the PLA now owns about twenty thousand companies and earns annual profits of around U.S.$5 billion.[5] According to some army sources, around 70 percent of officers' and soldiers' income can be traced to various corporations run by the PLA.[6]

Problems of Opportunity

Nevertheless, the aggressive participation of the PLA in business has increasingly led to serious problems, such as growing localism, smuggling, and corruption. For example, in the past decade, cases of smuggling and the illegal sale of Chinese relics to overseas merchants involving the army could not be thoroughly investigated by the legal authorities; the relevant departments in the army informed the legal departments that they would handle the cases themselves, but such cases were frequently not pursued or were merely handled internally. To attract potential business partners from abroad, some PLA executives even openly flaunted such privileges.[7]

In the mid-1980s, Chinese naval units stationed on Hainan Island colluded with local officials in smuggling more than U.S.$500 million in Japanese vehicles and television sets. The vehicles were later transferred to the mainland and resold at huge profits. Naval units have also been involved in smuggling along China's east coast. The most notorious case was the "Yellow Sea Smuggling," in which the North Sea Fleet ignored customs officials' interception efforts and caused severe damage by intentionally ramming customs vessels. In another case, PLA companies were allegedly accused of misdemeanors abroad, much to the embarrassment of the Chinese government.[8]

By the late 1980s, the PLA's involvement in illegal activities had become a headache for the government. The CMC leadership issued strict regulations called "Ten Nos" prohibiting PLA units from certain business activities. For example, PLA units were barred from using military vehicles or other resources for "speculation, profiteering or smuggling," and they were ordered not to "lend, hire out or sell military vehicle licenses, bank accounts or blank invoices."[9] In 1992 the Central Military Commission set up an audit commission in an attempt to enforce restraint.

In 1993, CMC leaders directed the PLA to pull out of some business activities nationwide but met with resistance. For instance, the army had been earning huge profits by exploiting its monopoly on mining and transportation in Datong, Shanxi province. The operations came to an abrupt halt when the PLA transferred its control of the twenty or so mines in the area to the provincial government. This "evacuation order" was issued directly from high in the army command.[10] Most units followed the directive, with the major exception of the 27th Army Corps, which refused to transfer control until it received substantial compensation. As a result, the central government was called in to mediate between the 27th Army Corps and the local government. The local government ultimately paid the division 1.3 billion yuan in compensation.[11]

In another move to curtail the army's business activities, CMC leaders decided in the same year to prohibit military personnel from simultaneously holding managerial positions as mid-rank company managers; the CMC also forbade army units to set up companies without permission from regional headquarters. The Central Military Commission further attempted to hinder the continued proliferation of PLA companies by consolidating its management at the group army level.[12]

Recently, serious concerns about the commercialization of the PLA have been expressed by both the party and top army leaders. Communist Party General Secretary Jiang Zemin, for example, warned that the PLA should concentrate on training rather than on business. CMC Vice Chairman Zhang Zhen clearly called on military districts to downsize their corporations and business activities.[13] Nonetheless, it will be difficult—if not entirely impossible—for the PLA to withdraw completely from the profit-making activities on which it has come to depend so much.

The Management System

In many ways, PLA companies may be defined as anomalous state-owned enterprises: They are by nature government enterprises but in reality are tightly controlled by the army. Many people continue to confuse PLA companies with China's defense industries. Although the two have a close relationship, they belong to two separate management systems—the PLA's companies are under the leadership of the Central Military Commission, whereas China's defense industry, like other civilian industries, is subordinate to the State Council (see Table 1.2).

The management system of PLA companies may profitably be studied at both the macro and micro levels. PLA companies fall roughly into three categories: national companies, regional companies, and group army companies. National companies refer to large conglomerates, which were among the first PLA companies to be established in the mid-1980s. Each general headquarters includes one or more of such conglomerates. The most well known include the Poly Group Corporation (*Baoli* in Chinese characters) and China Huitong Corporation (both under the General Staff Department); Carrie Corporation (*Kaili* in Chinese characters) and Tiancheng (both under the General Political Department); China Xinxing Group (General Logistics Department); China Xiaofeng Corporation, China Yuanwang Group Corporation, and China Xinshidai Corporation (COSTIND); China Lantian Industry Corporation (air force); and China Songhai Industrial and Commercial Corporation (navy).

All the military regions have also established their own comprehensive corporations. Among the most well known are Huabei Jinghai Enterprise Corporation (Beijing Military Region), Nanfang Industrial and Trading Corporation (Guangzhou Military Region), Dongbei Jincheng Industrial and Commercial Corporation (Shenyang Military Region), Northwest Industrial and Trading Corporation (Lanzhou Military Region), Huadong Industrial Corporation (Nanjing Military Region), Southwest Great Wall Economic Development General Corporation (Chengdu Military Region), and Shandong Dongyue Corporation (Jinan Military Region).

Group army companies have recently come under heavy pressure from the Central Military Commission for retrenchment. One decision made by the CMC in late 1993 was that all division-level and lower units would no longer be permitted to participate in business activities, although they could continue operating agricultural and sideline production. Group armies are still permitted to do business, but they are under stricter control. Those on the coastal areas tend to have more autonomous power; the 42nd Group Army's Changcheng (Great Wall) Huihua Industrial Corp, for example, currently operates more than ninety enterprises throughout Guangdong province.

The PLA business management system at the macro level is supposedly administered by the Production and Management Department (PMD) of the General Logistics Department (GLD). The PMD is an army corps–level unit headed by a major general and was originally created to oversee the PLA's own agricultural and industrial production activities within the GLD system. The PMD's official responsibilities include making long-term PLA business development plans, inspecting enterprises, managing human resource issues, and monitoring large production projects. Each military region also has a PMD located within its logistics department; it is responsible for direct management of PLA companies at the regional level.

Because of the lack of accessible reports on PMD activities, estimating the effectiveness of the PMD system in exercising control over the PLA's companies is difficult. According to various sources, however, the PMD has little control over national companies attached to the three general headquarters and insignificant influence on other national PLA companies. With regard to regional and group army companies, the PMD does exercise a fair amount of influence through the operational framework of the logistics departments of these army units.[14]

At the micro level, PLA companies resemble paramilitary organizations. Only very few top executives hold military ranks and most managers and employees are civilians, although many PLA companies still provide officer treatment to their senior civilian managers. Kaili, for example, provides benefits to its managers on the basis of their corresponding ranks in the army. Because Carrie is an army corps–level trading company, the president of Carrie holds the rank of major general and the rank of department heads corresponds to those of division leaders (such as colonel) in spite of the fact that they are civilians. Their ranks are used primarily to differentiate the compensation levels within the company.

Like other state-owned enterprises, PLA companies have an installed party system that ensures the political correctness of the operation of these companies and the leadership of the Party. However, the Party secretary, who holds the rank of a commissar, no longer has the traditionally dominant power. Normally, a vice president of the company represents the interests of the Party, but he or she also has the clear responsibility of assisting the president in rendering the company profitable. Because profit is the priority of these army companies, they increasingly resort to open recruitment of quality talent from outside the army and have appointed many university graduates to key positions of management. To be sure, many of these new appointees are sons and daughters of retired or current senior officers, but still the proportion of retired officers in these companies is shrinking.

In terms of formal organizational structure, these organizations mirror the highly centralized structure of state-owned enterprises. For instance, in the Poly Corporation, the president reports to a board of directors and is directly supported by a vice president, assistant president, and the general supervisor of finance, who in turn oversee a number of functional departments (see Table 10.1).[15]

On the informal side, PLA companies differentiate among three kinds of employees: army officers, semimilitary executives, and average employees who are recruited openly or semi-openly. The first two groups of people

Table 10.1 Poly Group Corporation Hierarchical Chart

Board of Directors

President

General Supervisor of Finance

Vice President

Assistant President

Departments

General Affairs

Auditing

Personnel

Finance

Policy Research

Planning

have access to internal information and are regarded as insiders. They generally enjoy lifetime employment (although army officers can be transferred or forced to retire from the companies). In contrast, the average employees are treated as outsiders and have little access to internal information. Their chances for promotion to top positions are slim, and they can be laid off whenever necessary.[16]

The leadership and decisionmaking styles are considerably more top-down and authoritarian than in similar kinds of civilian companies; frequently, the vital decisions are made by top executives with inside information. Employees of military organizations must take orders from superiors without questioning the validity of those orders. Many army companies were initially created by officers who had either blood ties or personal connections with senior leaders. The performance of these companies has been heavily dependent upon the personal power of these corporate CEOs, and the most successful PLA companies are known for the connections of their leaders.

The president of Poly, for example, is Major General He Ping, director of the Equipment Department of the PLA General Staff, which is responsible for liaison with COSTIND. More significantly, Major General He is also the son-in-law of Deng Xiaoping; his wife Deng Rong, the author of Deng's biography, is said to be Deng's favorite daughter. Another top figure in China Poly is Wang Jun, son of the late Wang Zhen, Long March veteran and former president of China. In addition to his position as chairman of China Poly, Wang Jun holds the title of president of China International Trust and Investment Corporation (CITIC).

The chairman and CEO of Carrie, Yue Feng (original name Ye Xuanning), heads the liaison department of the General Political Department. The second son of the deceased Marshal Ye Jianying, he is widely regarded as one of the most well-connected generals in the army. His elder brother, Ye Xuanping, is currently vice chairman of the Chinese People's Political Consultative Conference and is the former governor of

Guangdong province.[17] Partly due to Yue's personal influence, Carrie obtained many trading privileges during its initial period, including the power to trade with countries lacking official ties with China. Unlike Poly, Carrie does not deal in the weapons trade, although it has the authority to do so. In order to maintain a good international image, Yue Feng personally decreed that the company would ignore the weapons trade.[18]

Thriving in Market Jungles

One widely held misconception is that PLA companies engage mainly in weapons trading, such as missile sales to Iran and Pakistan, weapons transactions with developing countries, and the purchase of advanced weapon technologies from the West. In fact, most PLA companies are only marginally involved in such activities; their principal thrust of business is in the broad civilian market. Their participation is largely based on the assumption that they must be profitable. To achieve this sole aim, they have not restricted themselves to specific areas and, as mentioned earlier, operate in diversified business sectors.

In a sense, PLA companies are full-fledged participants in market competition and are often among the toughest competitors. Although there are losers in this competition, many PLA companies have excelled. Given their establishment during the mid-1980s, the speed of their expansion is remarkable. Within a short decade, major national PLA companies have developed into large conglomerates. This "miracle" has not been accomplished by a sophisticated management system; rather, these PLA companies have enjoyed a number of powerful competitive advantages, which are outlined in the following five points.[19]

Five Strengths

1. *PLA companies often hold monopolistic advantage in key business areas.* For example, the PLA controls an extensive transportation network, which is a major competitive advantage in an economy suffering from infrastructure bottlenecks. For years, PLA troops built many of China's airports, railways, harbors, and roads, originally meant for the PLA's exclusive use.

Since 1978, various PLA companies have turned these strategic projects into profitable assets. For instance, PLA companies played a dominant role in coal mine development in Shanxi province because they were among the few players that could actually transport coal to the industrially developed coastal and southern areas; local companies, lacking in transportation facilities, devoted themselves to coal extraction because they did not have the means to bring the coal to market.

In 1986, the air force launched China United Airlines, which operates commercial passenger and freight transport fleets from military airports. With more than twenty aging Tupolev and Boeing aircraft, it flies many domestic routes not normally served by civilian airlines. In 1992, it transported more than 400,000 passengers and 8 million tons of cargo. In a similar fashion, various navy units provide profitable shipping services from their own ports.[20]

In the telecommunications industry, the PLA shares oligopolistic influence with the Ministry of Post and Telecommunications and the Ministry of Electronics. China Huitong Corporation manages a nationwide telecommunications network. Thus, the PLA has become an attractive partner for many foreign telecommunications companies. Star Paging of Hong Kong, for example, made a deal with the PLA to develop one of its frequencies. Star plans to develop the largest cellular phone network in the country; it was selected partly because of its good connections with the army and partly because large international telecommunications firms like AT&T, Cable and Wireless, and Northern Telecom were hesitant to form partnerships with the PLA, which controls the majority of China's cellular frequencies.[21]

2. *Many PLA companies have benefited from advanced PLA research and development (R&D) institutes.* These technologically advanced facilities perform work ranging from space technology, machine tools, chemical engineering, and pharmaceutical manufacturing to agricultural production. Whereas many civilian companies suffer from an acute shortage of R&D funds for new products, PLA companies may take advantage of existing facilities and talent for new product development without making a heavy investment—it is paid through the military budget. The PLA maintains some of the best universities and research institutions in the country, such as the National Defense Science and Technology University in Changsha, the Shanghai Second Military Medical University, and the Longyang Foreign Language Institute.

The rise of the 999 Group clearly illustrates this advantage. Zhao Xinxian, 999's chairman and founder, was a professor of pharmacy in the army. In 1986 he founded the 999 drug group, which produces pills, powders, and ointments that claim to combat such common troubles as headaches, stomach problems, and skin disease. Unlike old-style Chinese medicine—herbs, roots, and animal parts, which require cooking and laborious preparation—the 999 Group medicines are sold in ready-to-use, Western-style packages. Sales reached 2.5 billion yuan (U.S.$288.1 million) in 1993, a significant jump from 60 million yuan in 1988. The 999 Group now controls thirty-four other companies in electronics, hotels, real estate, textiles, food, bottled water, export-import, investment, and construction.[22]

The Changsha Galaxy Technology Group, created by the Computer Department of the PLA National Defense Science and Technology University (PLA NDSTU), is another example: It has about three hundred senior engineers in the research area of computer architecture, system software, applications, networking, and multimedia.[23] Small as it is, the company has built the largest computer (in terms of capacity) in China—the Galaxy Series—and has successfully marketed Macintosh computers to the banking system of China. The company is now modernizing the operation system of the Bank of Industry and Commerce, one of the largest banks in the country. The company can easily draw on PLA NDSTU's Computer Department, which has the best hardware and talent in the field. Some of its managers are internationally known scholars in computer science. Gaining access to such facilities and skill has always been a daunting challenge for many civilian companies, but Galaxy can have both at almost no charge—both the facilities and personnel are paid through the military budget, through investment originally destined for military R&D. Such free rides for PLA companies are commonplace.

3. *Many army units have choice land, ready access to capital, and a plentiful labor force.* This gives them enormous advantages in a number of business operations, such as hotel and commercial real estate development and large civil construction projects. Many tracts of expensive land are part of army barracks and can be used as leverage in negotiating deals with developers. Because few banks seem to worry that the army will go bankrupt, they are generous in lending money to PLA companies. The PLA has its own professional construction troops for various military projects, and soldiers in nonconstruction troops can now participate in commercial projects. Many PLA companies have therefore utilized soldiers for commercial construction.

For example, some PLA-owned hotels are among the best located and constructed in the country, such as the five-star Palace Hotel in Beijing, one of the most expensive hotels in China. China's largest nightclub, J-J, is built on army-owned land in Shanghai. The 999 Group's construction company hires only PLA soldiers, aggressively bidding for projects not only in China but also in Macao. Its executives openly claim that nobody can underbid 999 Group. If worst comes to worst, and 999 Group loses money by underbidding, the PLA will pay the bill.[24] Without these advantages, many civilian companies simply cannot compete with PLA companies.

4. *PLA companies enjoy some special policy privileges.* Until recently, most PLA companies had paid either minimum taxes or no taxes at all. Paying taxes is a gray concept for many PLA companies, as they were established to compensate for the deficient military budget rather than to pay taxes to increase state revenue. Many PLA executives have stated in confidence that their companies paid no taxes. Moreover, in the absence of a feasible auditing system, it is impossible to establish the actual earnings of PLA companies. In one unexpected checkup raid conducted by the GLD in 1992, more than three hundred bank accounts with illegally retained money were discovered.[25] This catch only hinted at the extent of the problem; billions of yuan in earnings each year are not reported to the PLA authorities. Finally,

the relationship between many army units and their companies is never clearly defined. Taxation is almost impossible in the absence of a clear distinction between the PLA's military endeavors and its businesses. It is absurd for a government to tax its own army, but the taxation of its business remains a gray area.

There are other significant policy advantages as well. For instance, large PLA trading companies receive many import and export licenses. In fact, some companies obtain so many licenses each year that they simply cannot use them up, as one Kaili executive complained.[26] In contrast, many civilian import-export companies suffer from a chronic shortage of licenses. In order to conduct some of their import-export business, they must collaborate with the relevant PLA companies. This naturally places them in a disadvantageous position.

In addition, imports for use by the army are not subject to duties. Of course, it is often difficult to differentiate between use by the army itself to implement its military mission and use by the army for profitable gain. For the past decade, many PLA companies have taken advantage of this vast loophole for profitable activities. For a time, customs offices could not even check the contents of PLA imports, because opening these boxes was allegedly illegal.[27] In fact, it is common knowledge that most PLA imports have nothing to do with military applications.

Obviously these preferential policies constitute a tremendous advantage for PLA import-export companies. Not only are the costs of importing for these firms lower than for their civilian competitors, but also the processing time for importing is much shorter—perhaps up to half the duration. In sum, PLA transaction costs can be considerably lower than those of civilian firms.

5. *Furthermore, PLA companies are seldom harassed by local governments.* The widely held belief that PLA companies can deliver what they have promised and act quickly once they have accepted a deal is largely based on the extraordinary power PLA companies command at local levels. In fact, many PLA companies often enjoy such close relationships with local governments that they can cooperate for profit despite explicit directives from the central government and the Central Military Commission.

In this regard, the old saying that "the mighty dragon cannot defeat the local snake" does not apply. Many "local snakes" that try to find problems with PLA companies ultimately hurt themselves. At the same time, some central government–sponsored campaigns cannot touch PLA business establishments; for instance, at the height of antiprostitution campaigns in the early 1990s, local police tended to avoid raiding army-owned hotels. In some areas, this has become such an open secret that prostitutes flock to those army hotels that do not exercise control.[28] Even crooked companies in China have tended to be reluctant to cheat PLA companies, for there could be serious consequences and swift punishment if the PLA suspected any wrongdoing.

Three Weaknesses

The competitive advantages of the PLA may imply that PLA companies have no rivals. In spite of these advantages, however, PLA companies have some fundamental weaknesses. If these weaknesses are not well controlled, they may cause severe damage.

1. *As mentioned earlier, decisionmaking is a highly centralized process.* It occurs among a few, authoritarian, top executives, who may have vision and connections but lack well-developed management skills. This may work when companies are small or even medium-sized, but as firms grow into multinational conglomerates, this top-down style comes under increasing pressure. Because China lacks a tradition of decentralized management, change is not easy. Moreover, too much decentralization also means the loss of control by the army, which naturally hesitates to push for such change.

2. *Senior executives, who are also officers, are in a relatively tenuous position.* For the past decade, the PLA has rigorously implemented a program of keeping its office corps young. Consequently, PLA officers have a much shorter career life than do their counterparts in the government or elsewhere. Many must retire at their prime: A colonel will be forced to retire at the age of 55 if he is not promoted to the rank of major general, and a major general will have to retire at the age of 60 if he is not promoted to lieutenant general. Once retired, their positions in PLA companies also change correspondingly. They can no longer represent the army within the companies if they choose to stay, and their managerial power is considerably weakened. In order to create better opportunities for themselves after their retirement from the army, such executives tend to make hasty decisions on projects with not only high risks but also quick and substantial profits. Such short-term behavior has been very harmful to the steady development of some PLA companies.

3. *More significantly, the very nature of the PLA's close relationship with the Chinese Communist Party reveals a unique vulnerability of PLA companies to major changes in the macro political environment.* Top leaders are more sensitive to the PLA's business activities than to similar activities of other state-owned enterprises. Executives in PLA companies are fully aware of this vulnerability and have accordingly been opportunistic in many business transactions.

Under normal circumstances, a government would not allow its own army to be engaged in business. In a sense, the current situation resembles the one in Yan'an during the Guomindang's economic blockade, but the PLA's current financial dilemma is caused by none other than the Communist Party and the Chinese government. Nonetheless, the government shortfalls and stringency are abnormal, and once they are corrected, the rationale that led the army to operate its own businesses will disappear.

At the moment, no one seems to know when this day will come. This uncertainty has been very harmful to the long-term planning of many PLA companies.

The Formation of Diversified Conglomerates

Although most PLA companies are still young, they have already diversified their business activities. The majority began as trading companies for the relevant army units, then quickly expanded into general trading companies for nonarmy clients, and finally branched into investment and manufacturing businesses. Within a decade, several national PLA companies have grown into the most diversified conglomerates in the country. The speed of their growth and the scope of their diversification can be matched only by the Korean *chaebols*.[29] Xinxing, Poly, and Kaili, which were created during the mid-1980s and formally approved as business groups by the State Council in 1991 and 1992, serve as good examples.

Xinxing officially owns about one hundred subsidiaries, which in turn have many of their own ventures. China Xinxing Import and Export Corp (CXXC) is the core of China Xinxing Group. With branches in major Chinese ports, CXXC has been the sole exporter of military supplies and logistics equipment, along with exports and imports of cereals, oils, foodstuffs, native produce and animal by-products, light industrial products, textiles, handcrafted articles, metals, minerals, medicines, health products, machinery tools, instruments, and meters. CXXC also engages in barter trade, coproduction, and processing ventures. Most of its current business activities have nothing to do with the PLA. In 1992, the first year that Xinxing became a conglomerate, the Xinxing Group revenue was estimated at 7 to 7.5 billion yuan (more than U.S.$850 million), with about one-tenth representing profits.[30]

As with Xinxing's trading operations, Poly's import and export operation has been its leading business, but on a much larger scale—an annual import and export value of U.S.$500 million.[31] Originally based on defense equipment trade, Poly has quickly diversified into a wide range of civilian products. Today, Poly's products are a match for those of any civilian trading firm. Another major business is property and construction; more than twenty large and medium-sized development projects have been undertaken with Poly's investment in Beijing, Tianjin, Shanghai, Guangzhou, and other large cities as well as special economic zones. Among the most well known are Legend Garden Villas in Beijing, Hainan Yangpu Land Development, Shanghai Securities Exchange Building in Pudong, and Tianhe Plaza in Guangzhou. Poly has also been heavily involved in industrial investment, with such major investment projects as Hainan Yangpu Power Plant, Hebei Chongli Steel Plant, Three Gorges Mechanized Construction Company and projects in telecommunications, engineering, shipping, and even garment manufacturing.

Currently, Poly is expanding into the following service sectors: tourism, office-building leasing, restaurants, shopping centers, hotel management, entertainment, advertising, transportation, futures trading, and photo studios.[32] From its inception, Poly registered as a subsidiary of CITIC in order to deemphasize its military affiliation, although the linkage between the two corporate giants is merely nominal. In short, Poly Group is truly diversified; one of its executives once indicated that more than 80 percent of Poly's business is extraneous to the PLA.[33] If one examines Poly's business portfolio, one would be hard pressed to guess that Poly is a General Staff–owned conglomerate.

The smallest of the three highly diversified PLA companies, Carrie Enterprises, has followed a similar pattern of development. Headquartered in Beijing, Kaili has branches in Dalian, Tianjin, Jinan, Nanjing, Shanghai, Fuzhou, Guangzhou, Shenzhen, Haikou, Kunming, Wuhan, and Lanzhou, among others. At present, Kaili officially has fifty-six subsidiaries, but many of its smaller subsidiaries overseas are not publicly identified with Kaili. Some of them use entirely different names, such as Lida (Shanghai) and Liyuan (Nanjing). The annual trade income of Kaili ranges from U.S.$300–$400 million.[34]

Like Xinxing and Poly, Kaili trades in a variety of products, ranging from cereals and oils to high-tech products. In addition to regular trade, entrepôt trade, and various forms of counter-trade, Kaili has been quite active in operating joint ventures and cooperative production projects, paying special attention to industrial development. It has established sizable economic entities or enterprises in the fields of contractual international engineering projects, labor cooperation, real estate, audiovisual products, book publishing, energy, electronics, paper making, leather processing, and tourism.[35]

Kaili now owns the second largest paper factory in Asia, the First Paper Factory in Yantai, Shandong province. Carrie also owns a highly regarded leather factory in Xinjiang, which was established with aid money of U.S.$50–$60 million from the Spanish government. In addition, it possesses a printing plant in Fuzhou, a cloth factory and fashion store in Beijing, and a few successfully run power stations.[36]

Rapid Internationalization

Although internationalization is still a relatively new phenomenon for most Chinese companies, their interest in gaining a foothold in other countries, especially in the West, is very high. In spite of various governmental restrictions, many Chinese companies have managed to establish a presence outside China. Among the most eager pioneers are PLA companies.

With the imminence of Hong Kong's 1997 return to China, the PLA's business presence in the territory is expected to grow significantly. The GLD has already established a logistics base in Shenzhen to prepare for a PLA garrison in Hong Kong after 1997. Many expect that the garrison and its auxiliary units will be heavily involved in business activities in order to support themselves.[37]

Xinxing was among the first PLA companies to establish an open presence in Hong Kong by setting up a trading office in November 1992. The deputy director of the General Logistics Department, Lieutenant General Zhang Bin, flew to Hong Kong to attend the inaugural ceremony, which was also attended by a group of Hong Kong tycoons, including Li Kashing (Hong Kong investment tycoon and personal friend of Deng Xiaoping and senior leaders in China) and T. T. Tsui (Hong Kong and London business tycoon with longstanding PLA connections).

Other PLA companies have expanded into Hong Kong more discreetly through front companies. Instead of using its own name, Poly set up a trading company known as Ringo Trading in the territory in 1985. Reorganized in 1992, this company possessed starting capital of Hong Kong (HK) $50 million and has since diversified its business activities to include investments in property, shipping, and infrastructure projects in Hong Kong and China. Among its major projects and joint ventures are property developments in Shanghai, Fuzhou, and Hainan provinces involving Chinese Estates Holding and Peregrine Investments Company; the joint development of the Hainan Yangpu Economic Development Zone with Kumagai Gumi (Hong Kong) and Cheung Kong (Holdings); and the acquisition of listed companies in Hong Kong.[38]

In August 1987, Poly set up its first subsidiary in the United States—Dynasty Holding Company, located in Atlanta, Georgia. Its principal business covers import and export trade, property dealing, and management. Poly has also established a presence in such countries as Russia, Thailand, Myanmar, and Pakistan and business relations with more than two hundred foreign companies from the United States, Germany, Russia, United Kingdom, France, Singapore, Switzerland, and Australia.

The aggressive expansion into the United States by the 999 Group in 1994 made headlines in major U.S. newspapers. At the time, the 999 Group was bringing in products through San Francisco, preparing a factory in Florida, negotiating an R&D venture in San Diego, and planning to list its shares on the New York Stock Exchange. Such moves were reported in such a dramatic way that concern was raised among many in the country. The U.S. House of Representatives debated a proposal by Representative Nancy Pelosi (D-Calif.) to restrict imports from PLA-linked Chinese firms such as the 999 Group.[39]

Nevertheless, the 999 Group had already lined up some powerful U.S. backers. An Asian unit of Ciffcorp has acquired a 2.4 percent stake in the company. Two other U.S. investment funds, Paloma Partners of Connecticut and Paul Ibdor Jones's Tudor Investments Group of New York, directly and indirectly control 25 percent of the 999 Group's shares, which are not publicly listed. Despite concerns expressed by many in the U.S. Congress over 999's business expansion into the country, Pelosi's proposal was rejected.[40]

Although other major PLA conglomerates have received more attention in their aggressive drive for internationalization, Kaili is probably the most international of PLA companies. By 1995, Kaili had established an international empire, but very few of its subsidiaries abroad openly acknowledge the relationship. In general, most of its subsidiaries are relatively small operations, unknown to many in the business. Even people working in Kaili have limited knowledge of the existence of these subsidiaries and their activities. Many of these companies were allegedly set up to perform a very diversified mission in addition to simple business internationalization, although this contention has not been substantiated. To facilitate its internationalization, however, Kaili has somehow obtained certain privileges, including a special quota for emigration from the mainland to Hong Kong and permission for its expatriates to seek permanent residency in the host countries in which they work. In this regard, Kaili has probably the most internationalized staff; its members can mingle well with the citizens of host countries and many simply pass as locals.[41]

On the whole, the rationale for the internationalization of PLA companies is founded on risk reduction through diversification in terms of business and location. In so doing, these firms also learn international practices and can transfer this knowledge back to China. However, internationalization also reduces interference from the central government or the Central Military Commission and allows these firms to make reverse investments in China in order to gain privileges as "foreign investors."

For their own protection, many PLA overseas subsidiaries have either maintained a low profile or simply used different identities that do not associate them with units of the PLA. Therefore, concerns over the political implications of these PLA international activities are somewhat exaggerated; thus far, most of these companies are primarily in the business of making money rather than extending the political interests of the Communist leadership beyond Chinese territory.

At the same time, one cannot totally ignore the impact of political or power struggles within the top Chinese leadership upon PLA companies. As has already been mentioned, unlike most other Chinese state-owned companies, they are highly vulnerable to the political manipulations of Communist leaders. In spite of the continued marketization and reform of more than fifteen years, the PLA must still follow the long-established principle that the Party leads the army. This has been the iron rule of their relationship since it was laid down by Mao Zedong during the Long March in the mid-1930s. The PLA was, is, and will in the foreseeable future be the power tool of the

Party: It has always been the army of the Party. In this sense, the rapid internationalization of PLA companies may have future political ramifications.

Conclusion

Despite growing concern in the Chinese leadership over PLA business activities, no clear resolution of the many issues raised by those activities has been reached. Indeed, the Chinese leadership faces a catch-22 situation. Prohibiting the PLA from participating in business will have immediate consequences for the financially strapped central government; it would also give rise to heightened bitterness among many officers, who already feel abandoned by the reforms and who resent the loss of their previous levels of prestige and respect in society. With the Deng Xiaoping era drawing to a close, any such feelings among PLA officers are the last thing that the current leaders need in this period of uncertain power transition.

However, the uncontrolled business activities of PLA companies may weaken the army's professionalism and worsen corruption among the officer corps. This could lead to rising social tensions and heightened bitterness and bad feelings among the Chinese people toward the army. Moreover, continued marketization requires a minimal basis of fair competition, but the existence of a large number of specially privileged PLA companies could thwart further marketization and distort fair competition. Even worse, an army with enough of its own financial power could disregard the leadership of the Party and the government, thus weakening the Party's control.

Thus far, no solution seems to be the best solution. The most likely trend of development is the corporatization of PLA companies, which will eventually turn army conglomerates and other companies into limited-liability corporations—as we have seen in the civilian-run defense sector. This may be the most constructive and effective way to gradually separate the PLA from its own companies. Once their assets are objectively appraised and their ownership is clarified, PLA companies will be able to issue shares to both their military units and other sectors of society. In due course, they will develop their own identity and become truly independent business organizations. Market competition will also force many small PLA companies to either consolidate into large conglomerates or join existing conglomerates. The process of consolidation and merger may help PLA leaders rid themselves of those many unwanted and worrisome "guerrilla forces" that have created most of the scandals. In this possible future the process of the development of PLA companies will be a slow but uniquely Chinese way of converting military resources to civilian benefit.

Notes

1. Documented information on this topic is scarce, and the available information is often unreliable. Much of the information in this paper was obtained through the author's private interviews with PLA business executives. The author wishes to express his gratitude to these PLA executives for their contributions.

2. Wang Fang, "Fighting at Wartime, and Pursuing Business at Peace Time," *China Daily Magazine* 188 (6–12 August 1995), p. 16.

3. The "Big Production Campaign" has been covered in detail by history books of the Chinese Communist Party but has been treated only superficially in the West.

4. Interview with a senior Kaili executive, Beijing, March 1995.

5. "China's New Model Army," *Economist,* 11 June 1994, p. 31.

6. Based on interviews with a number of PLA executives, Beijing, March 1995.

7. Based on discussions with a number of PLA executives, Shenzhen, July 1994.

8. Wang, "Fighting," p. 18.

9. Tai Ming Cheung, "Serve the People," *Far East Economic Review,* 14 October 1993, p. 66.

10. "China's Enterprising Army," *World Press Review,* January 1995, p. 36.

11. Wang, "Fighting," p. 18.

12. Based on interviews with executives from a number of PLA companies, Beijing, July 1994.

13. Ibid.

14. Based on interviews with a number of PLA executives, Shenzhen, July 1995.

15. China Poly Group Corporation, *Annual Report* 1993, Beijing, p. 11.

16. Based on interviews with many PLA executives, Beijing, 1994–1995.

17. For a brief introduction to the Ye family, see Carolyn Koo, "The Princelings of South China," *Institutional Investor* 29:5 (May 1995), pp. 103–106.

18. Interview with a senior Kaili executive, Beijing, March 1995.

19. Min Chen and Wangjiang Ying, "Learn How to Deal with PLA Corporations," *Zhong Shi Zhou Kan* [China Times Weekly] 106 (9–15 January 1994), p. 33.

20. Cheung, "Serve the People," p. 65.

21. "In with the Army," *Business China* 20:12 (13 June 1994), p. 12.

22. E. S. Browning, "China Drug Company with Ties to Military Is Expanding to U.S.," *Wall Street Journal,* 7 September 1994, p. 7.

23. Interview with executives of Changsha Galaxy Technology, Changsha, August 1995.

24. Interview with an executive from the 999 Group, Hong Kong, July 1995.

25. Wang Yihua, "Preliminary Achievement Made in All-Army Outlay Final Accounting," *People's Daily,* 31 May 1992.

26. Interview with an executive from Kaili, Beijing, March 1995.

27. Interview with a PLA business executive, Beijing, October 1995.

28. Based on discussions with several PLA business executives, Shenzhen, July 1995.

29. Dong Sung Cho, "Diversification Strategy of Korean Firms," in *Korean Managerial Dynamics,* eds. Kae H. Chung and Hak Chong Lee (New York: Praeger, 1989), pp. 99–112.

30. Based on interview with a trading executive from Xinxing, Beijing, July 1994.

31. China Poly Group Corporation, *Annual Report* 1993, p. 9.

32. Its headquarters, Poly Plaza, covers an area of 20,000 square meters with a construction area of 87,000 square meters. It offers a 1,230-seat modernized theater, a 500-seat multifunctional hall, an exhibition hall of 12,000 square meters, 280 sets of deluxe rooms, luxurious restaurants, a large swimming pool, health center, beauty parlor, salon and sauna, and other modern facilities. Interview with a Poly executive, Beijing, July 1995.

33. Ibid.

34. Interview with a Kaili executive, Beijing, July 1995.

35. China Kaili Enterprises, company brochure, Beijing, 1993.

36. Interview with a Kaili executive, Beijing, July 1995.

37. Cheung, "Serve the People," p. 66.

38. China Poly Group Corporation, *Annual Report* 1993, p. 47.

39. Browning, "China Drug Company," p. 7.

40. Ibid.

41. Interview with a Kaili executive, Beijing, March 1995.

PART 5

Concluding Perspectives

11

Semantic Differences: Comparing Defense Conversion in China and Russia

Jörn Brömmelhörster

Given that conversion is an interdependent process initiated by both domestic and foreign influences, it seems necessary to examine the experiences of conversion in other countries to obtain a broader understanding of the concept. The form and the extent of domestic conversion efforts depend to a large degree on the success or failure of conversion in neighboring and/or antagonistic countries. It is therefore useful to compare the experience of conversion in Russia with that in China.

The two countries have tight historical links. Since the rise of the People's Republic of China (PRC), the Soviet Union turned from a friend and partner in development to a foe and an ideological enemy, with a sudden reversal toward cooperative endeavors in recent years. Moreover, both countries share one of the longest borders in the world, with very high troop concentrations still present on both sides. The size of both countries also accents their comparability, and both are major players in world politics as well as in the global economy. From the political point of view, they both developed their own socialist philosophy and, in recent years, distanced themselves from its original form and even from its very substance. This evolution has led to dramatic social and economic transformations within each country. Last but not least, they have both repeatedly expressed in official documents their desire to permit and enhance conversion activities.

The first section below provides information on similarities and differences in political and economic conditions and tries to clarify the impetus for conversion in the two countries. The second part deals with policy, specifically highlighting the approach to conversion, its various stages, the interrelations of these stages, and its effectiveness.

In the third section, initial successes and residual hurdles to conversion are discussed. In this context, three major economic factors have been chosen that ultimately determine the success or failure of the conversion process: first, geographical location and concentration of arms-producing factories; second, the extent to which the readjustment of human resources will slow down or accelerate the conversion process; and finally, the need to take into account the means of companies to finance conversion activities and the resulting consequences for the companies themselves as well as the overall economy. Apart from this economic discussion, the military-industrial complex in the respective countries and its adverse or positive impact on conversion is compared and analyzed. In conclusion, I seek to consider whether or not the policy of conversion will be pursued based upon the previously mentioned economic and institutional factors.

Initial Positions

Preliminary measures toward the proclamation of conversion were taken in China one decade earlier than in the Soviet Union. While antagonism between the Soviet Union and the United States had been leading to significant modernization of their arsenals and therefore to an increase in real terms of their military expenditures—by 6 percent for the USSR and more than 40 percent for the United States between 1980 and 1985—the military expenditure of the People's Republic of China only rose by 1.7 percent in that period.[1] During the armaments race between the Warsaw Pact and NATO, China stood at the beginning of the Deng Xiaoping era, which was characterized by a radical break with Maoist principles. The autarchy efforts that had predominated under Mao were abandoned in the late 1970s, giving way to a phase of "opening" policy and to the "Four Modernizations," the fourth of which—the last on the priority list—was the modernization of national defense.[2]

The first conversion phase in the USSR was launched by then-President Mikhail Gorbachev's programmatic speech before the United Nations in December 1988, in which he announced unilateral reductions of the Soviet army.[3] Consequently, conversion became an important part of perestroika. In the USSR, glasnost—which involved political opening—was integrated into the general transformation strategy, whereas this "fifth modernization" was rejected in China.[4]

A brief look at the two countries' geopolitical situations at the initiation of conversion shows that China was at war with Vietnam at the end of the 1970s and that external conditions were not favorable to conversion. Likewise, the Soviet Union in the mid-1980s was involved in the arms race with the United States (the two military blocks, NATO and the Warsaw Pact, accounted for approximately 80 percent of worldwide defense expenditure at that time) and was still fighting in Afghanistan.

Table 11.1 indicates that in the initial stages of conversion, the two countries were at different levels of development. In terms of gross national product (GNP) per capita, China in 1978 was one of the poorest countries in the world, whereas the Soviet Union was counted in the group of the twenty-five richest countries in 1988. Both countries followed socialist

Table 11.1 Indicators at the Beginning of Conversion Implementation

	Military Expenditure as Percent of GNP[a]	Soldiers	Employees in Defense Industries	Ratio of Heavy to Light Industry	GNP Per Capita (in U.S. dollars)
China (1978)	9.6	4,500,000 (4.6)	4,000,000 (4.2)	57.3/42.7	465
USSR (1988)	12.6	3,900,000 (13.6)	6,300,000 (22.0)	74.8/25.2	9,180

Sources: U.S. Arms Control and Disarmament Agency (ACDA) data; Bonn International Center for Conversion data; *Zhongguo Tongji Nianjian* [Statistical Yearbook of China] (Beijing: Zhongguo Tongji Chubanshe, 1985), p. 309; *Narodnoye Khozyaistro SSRV* [Economic Yearbook of the USSR] 1989 (Moscow, 1990), p. 333.

Notes: a. Although some databases on military expenditure exist, it is very difficult to derive the real picture in terms of military dependencies, especially for the aforementioned countries. Due to different accounting methods, different definitions, and a nearly unimaginable propensity toward secrecy, the data should be understood more as a trend than as actual results. For a discussion on the data problem, see Michael Brzoska, "World Military Expenditure," in *Handbook of Defense Economics,* vol. 1, ed. Keith Hartley and Todd Sandler (Amsterdam: Elsevier Science B.V., 1995), pp. 45–67.

b. Numbers in parentheses are soldiers/employees per 1,000 people.

theories according to which heavy industry was developed as a priority. Thus, in both countries, the figures reveal the relative insignificance of light industry—in other words, the considerable need to catch up in the consumer goods industry. The USSR had placed an even higher priority on heavy industry and therefore had a relatively greater need to reach a more appropriate distribution.

With regard to the degree of militarization, the table shows that the defense expenditures of the Soviet Union accounted for 12.6 percent of GNP (according to U.S. Arms Control and Disarmament Agency [ACDA] data). After the dissolution of the USSR, however, new appraisals were carried out that led to much higher results, due to reasons of comparability. According to other estimates, military expenditure would have reached 20–25 percent of GNP.[5] The proportion of defense expenditure in GNP was much smaller in China than in the USSR, although the Soviet example may give reason to suppose that a new assessment of Chinese spending would result in a higher military dependency. In terms of personnel, the two countries have the largest armies in the world; however, the ratio of armed forces to the general population was much higher in the USSR (13.6 per 1,000 people) than in China (4.6 per 1,000 people). These data can only be interpreted as a trend in order to draw conclusions on militarization; they do not take into account that the "people's war" doctrine implemented in China at that time resulted in a much more extensive incorporation of the population through military training and the formation of militias.[6]

The high defense burden on the Soviet Union is also documented by the fact that out of one thousand people, twenty-two worked as defense employees.[7] The Soviet Union produced advanced and state-of-the-art weaponry, whereas Chinese production of military goods lagged at least twenty years behind. This discrepancy may be explained not only by China's lower level of development, but also by its implementation of a military doctrine according to which the masses of people outweighed the quality of technical equipment.

In sum, conversion in China was one of the results of internal political changes caused by the abandonment of Maoist principles and the opening toward economic restructuring rather than a reaction to high defense outlays and external factors. In contrast, the pressure—but not necessarily the necessity—to implement conversion in the Soviet Union resulted from the arms race, which had caused a nearly unbearable defense burden.[8]

Stages of Conversion Policies

China

In considering the different conversion stages in China, one is struck by the fact that conversion emerged as a political necessity. Defense orders were canceled due to policy changes, and the emphasis simultaneously shifted to economic construction. The defense industry therefore had to adapt. During the first stage, from 1979 to 1982, companies had to fend for themselves; however, they had the right to penetrate into the civilian sector, and adaptation strategies were implemented outside the military sector as well. Reductions in the military-industrial complex (MIC)—which were allowed despite the campaign in Vietnam, the lasting conflict with Taiwan, and the threat of a better-equipped, China-targeted Soviet army[9]—point to the presence of surplus capacities and resource waste.[10]

The victims of this new orientation were companies producing less important defense goods and those in remote ("Third Line") locations.[11] Nonetheless, they were victims only in the sense that they produced even fewer goods than before. Belonging to the defense sector provided a certain security—cuts in payrolls were not expected, and only a few, partial closures were reported.[12] This may be explained by the fact that, in the beginning, only short-term reductions were anticipated and the typical surge capacity in the defense sector was taken into account.

In contrast, the more modern and efficient companies should have been able to take advantage of their potential during this stage; nevertheless, conclusions drawn retrospectively about this stage claim that it had little success. The isolation and the market ignorance of the defense industry generated almost no marketable products. Of course, one must question how new product lines could have been developed in such a short time. Relatively successful companies were those that either produced dual-use goods or had the opportunity to invest capital in new civilian businesses. In most companies, the goods that could be sold had been produced at higher costs than comparable civilian goods.

The first stage may be understood as a period of trial and error (see Table 11.2). Procurement had been cut, and the defense industry was forced to react. Subsequently, it had to be decided whether or not "spontaneous production" (*zifa shengchan*), the Chinese designation of this production stage, had borne fruit. With regard to the implementation of conversion understood as the production of marketable civilian goods, it achieved almost no success. This stage was not completely useless, however, because it provided the opportunity to discover the location of specific weak points of the defense sector.

Table 11.2 Periods of Conversion

China	Russia (Soviet Union until 1991)
Trial and Error (1979–1982) The four modernizations result in a shift of priorities toward economic development. Cuts in procurement lead to experimentation with conversion goods.	*Command Defense Conversion (1988–1991)* Gosplan launches the first conversion program to be implemented within the command economy, with its inefficiencies, and the traditional structure of the MIC.
Definition Period (1982–1985) The "16-Character Slogan" sheds light on conversion. It is viewed as a long-term strategy to serve national economic construction. Campaigns promote defense conversion.	*Conversion Act Period (1992)* A strategic framework exists without sufficient financial means in the context of a contradictory economic policy.
Reorganizational Period (1986–1991) After the new content, a new structure is created. It includes stronger but insufficient and hardly stimulating government measures.	*Conversion Program Under Budget Contraints (1993–1995)* An ambitious program with priority sectors raises great expectations but fails due to undercapitalization.
Business Focus (since 1992) Strategies are generated for reorganizing defense conversion companies. They include some incentives but many conflicts.	*Conversion with Less Government (since 1995)* A qualitative and quantitative shift of public conversion strategy occurs, involving the definition of what should remain.

The second stage began in 1982 when Deng Xiaoping gave a precise definition of conversion, the "16-Character Slogan": "Civil-military integration, integration of peacetime construction with war preparedness, priority to military production, and support the military with civilian production."[13] The objective extended beyond the desire for both guns and butter;

the butter should contribute to financing the production of guns. One year later, in 1983, emphasis was placed on the fact that the defense industry's "butter" constituted a long-term support for economic construction.[14] Cumulatively, these points amounted to a strategic reorientation to a "peacetime strategy," which took the new security environment and the abandonment of old Maoist doctrine into account. Therefore, investments to allow for limited retaliation against nuclear attacks could be lowered and portions of the army could be demobilized.[15]

Conferences and fairs were organized to market conversion goods. Industries from remote regions had the right to open sales offices and partial production sites in the coastal regions to improve marketing and increase production efficiency.[16] The goods presented at the fairs clearly show how similar existing notions of conversion goods were in the defense industries. In particular, the majority of companies tried to enter or expand into the consumer goods market (television sets, refrigerators, fans, motorbikes, and so on). Entering or expanding into the consumer goods market may have been the correct strategy—China had been characterized by excessive purchasing power and a simultaneous scarcity of consumer goods. However, this strategy will work only when marketable quality products are manufactured or when timely, limited import substitution is carried out at the same time.

The effect of this push to make consumer goods was an increase of the share of civilian goods in the defense industry from some 8.1 percent in 1979 to more than 40 percent by 1985.[17] The published figures of individual companies generally document an increase in production or sales revenue but do not furnish sufficient information on the potential increases in profitability. The literal application of Deng's "16-Character Slogan," moreover, suggests that rather unprofitable companies and surplus factories were spun off in the conversion efforts.[18]

In 1986, when the third period started, international tension remained high. Global military expenditures were still increasing; in addition, announcements by Gorbachev had made it clear that major policy changes would occur. From the economic point of view, the policy of opening showed greater success, and instruments that had proven their effectiveness in experimental trials in the special economic zones (such as tax benefits for foreign investors) were extended—in small doses—to other regions. From the institutional point of view, this stage saw the restructuring of former defense ministries, which thereafter were assigned to the State Council.[19] This restructuring had two effects: first, the strict separation between civilian goods manufacturers from the military sector and those from the civilian sector was abolished; second, this civilian production was directly subordinated to the national plan. This subordination to the national plan allowed increased integration with the civilian economy, toward better coordination ability. The objective was to increase efficiency by putting an end to the situation in which converting industries compete with civilian industries for the same customer.[20] The creation of the China Association for Peaceful Use of Military-Industrial Technology (CAPUMIT), which was subordinated to the Commission on Science, Technology and Industry for National Defense (COSTIND),[21] was a further step in this direction.[22] CAPUMIT promotes conversion projects and, as a frequent organizer of conversion conferences, establishes links with foreign investors to persuade them to join Chinese conversion projects. The improved institutional embedding also made conversion projects easier to prepare. In particular, two programs help support conversion through loans (approximately 4 billion yuan for the Seventh Five-Year Plan in 1986–1990, and approximately 6.3 billion yuan for the Eighth Five-Year Plan).[23]

The patriarch Deng's trip to southern China in 1992, which recalled the famous trip of Qing emperor Kangxi, was also a milestone for conversion. Subsequently, reform focused more strongly on the business and management level. Although previous signs of this strategy had existed, it was implemented on a broader level after 1992. Conversion would be carried out through enterprise reform; models such as the stock share system, the individualized management system, and the modern enterprise system took shape.[24] Although the consequences of this policy cannot be fully assessed yet, it has become apparent—especially in companies in which the civilian part is more successful—that such companies hesitate to accept military orders (because they signify losses) or do not accept them at all.[25] The gap between profitable and nonprofitable companies has become both clearer and wider. If the reluctance to accept military orders increases, new signals from the top can be expected; such signals might give more precise directions and exclude some companies—those that are important for military production—from implementing some of these new management systems.

According to figures for this period, about 80 percent of goods produced by defense industries are now civilian goods, and two-thirds of defense industries no longer produce defense goods.[26] The most impressive achievement is reflected not in these ambiguous figures but rather in the substantial number of new companies that have begun to manufacture civilian goods.[27] The conversion branch of the MIC involves a growing number of companies that are also considered crucial to maintaining the military-industrial base.[28] Even some cities that were erected or expanded for defense purposes, such as Mianyang and Panzhihua in Sichuan, have been opening to civilian activities.

To summarize, a shift toward conversion has occurred, but the economic outcome is unimpressive. The spectrum of loss-making defense companies lies between 20 and 50 percent, according to various sources.[29]

Russia

In 1989–1990, after the aforementioned pathbreaking speech of Gorbachev at the United Nations, the State Planning Commission (Gosplan) introduced a program aimed at increasing the MIC's civilian output from 40 to 60 percent by 1995 (see Table 11.2).[30] A priority list was established to reach this goal. The most important sectors included food processing and agricultural machinery, equipment for light industry, trade and public catering, consumer durables, electronics, computers, medical equipment, communications, civil aviation, and commercial shipbuilding. In addition, existing facilities would receive substantial new investment funds and be partially relieved of their military obligations. The majority of these items never reached the production stage because they either did not meet plan requirements or had been designed without any cost restrictions. Furthermore, by the end of 1991 most enterprises had run out of funds to continue such projects. It became clear that "command defense conversion" could simply not keep pace with drastic reductions in military expenditures forced by the deteriorating economy. The disintegration of the Soviet Union and a shift in the path of reform ended this phase.

The second stage of conversion commenced at the beginning of 1992, lasted only one year, and was characterized by a lack of state initiative. The "Conversion Act" passed by the Russian Parliament included a general strategy for dealing with the economic and social consequences of demilitarization but lacked any funding to support converting companies. Nevertheless, this strategy contributed to the decentralization of decisionmaking and the delegation of autonomy to individual enterprises. During this period, state subsidies and credits were used primarily to meet payroll requirements and therefore to ensure social peace rather than to implement conversion projects.

It was during this stage that a new concept arose, that of "economic conversion."[31] Overly optimistic expectations of arms sales on the order of U.S.$30 billion annually had led some government officials and defense managers to think that money for conversion projects could be earned by increasing arms exports. As a result, some enterprises began to produce weapons at levels above contractual obligations, and the government ordered more to meet the expected demand. Therefore, the production of weaponry mobilized massive amounts of resources; this excessive production, which could not be absorbed by domestic and international demand, led to stockpiling. Not only did such policies counteract conversion efforts, but the lack of clear signals concerning domestic procurement also discredited the concept.

The third stage of conversion once again focused on a priority list, one very similar to that in the first stage. The implementation of this ambitious program confronted major problems: the fundamental question of the efficiency of state investment allocations and priority setting and the expectations generated by such allocations. The announced funds either were not fully released or were considerably reduced by near hyperinflation. In addition, the scope of the program was misinterpreted as a continuation of state intervention.

In 1995, a fourth phase began that may be characterized as "conversion with less government." The government reduced its involvement in terms of size and strategy. Private investment was promoted in the sense that the state decided (or was forced) to limit its share in conversion projects to one-fourth. The limitation of state subsidies and the poor prospects of earning money from arms exports in most armament sectors resulted in more requests for bank loans so as to at least meet immediate payroll requirements. A new guideline was established, promising more planning security for the whole defense sector. Three categories of companies were defined:

- About 25 percent of arms-producing companies gained the status of "treasury" enterprises (*kazenniye predpriyatiya*). These companies work only for the military sector and remain state owned. Their management is subordinate to direct state control, they are paid directly by budget funds, and they work according to governmental plans.
- A further 25 percent also remained state-owned, or majority-owned by the state. They produce military as well as civilian goods. The companies work on a contractual basis with the government; managers are less dependent than those in the first category, but still serve upon approval by state authorities.
- The remaining 50 percent were fully released for privatization and conversion. They are not considered vital to national security but can nevertheless win military contracts, mainly as subcontractors.

This fourth phase has definitely clarified the situation in Russia. The state has withdrawn from much of the defense sector in determined boundaries—it retains only the vital sectors. Given the current unprofitability of the former arms-producing companies, however, it is doubtful whether 50 percent of them can be privatized without supplementary costs.

Policy Findings

In sum, several principal differences exist between the two countries. Russia features a much higher defense burden that must be reduced. China's defense outlays are by far lower than Russia's but must be restructured. The Russian arms industry had from the beginning a much stronger civilian orientation than did the Chinese one; the aforementioned 40 percent civilian share in the armaments sector was composed primarily of consumer goods (such as television sets, radios, video recorders, sewing machines, and refrigerators).[32] However, only a few companies have been able to draw advantages from these products because requirements and quality could not meet world standards.

In China, the civilian share was much smaller. However, consumer goods were almost unknown to the Chinese population. Due to this immense sellers' market, some companies could draw advantages from civilian production more easily (especially for motorbikes, vans, and television sets), although not all succeeded.

The different stages demonstrate that China uses the same pragmatic approach for its conversion policy as it does for reform policy: first, trial and error; if a strategy works, it is continued. Initially, trial and error policies are always carried out in limited sectors or regions. In addition, conversion has been interpreted broadly and is understood as both general reengineering and a way to ensure surge capacities.

Russia's conversion policy is also pragmatic, though in a different way: It tries to obtain immediate and widespread results, and when these fail, policy changes occur that can be best described as "muddling through" rather than as clear overall policy management. In retrospect, the different stages of conversion in Russia point less to crisis management in the defense industry than to a kind of chaos management. Many problems have been magnified rather than reduced due to contravening procedures and halfhearted action.

Achievements of and Remaining Barriers to Conversion

The description of the different stages of conversion showed that both countries initiated conversion measures but that their respective conceptions of conversion are very different. The following section analyzes whether these conversion efforts have borne fruit. If positive results were recorded, how stable is the trend and what barriers must still be overcome?

Economic Factors

The data used for Figures 11.1 and 11.2 clearly indicate that the ratios of military expenditures (milex) to GNP and to central government expenditures (CGE) have been decreasing in both the USSR and China.[33] Since 1991 the ratio of milex to GNP increased again in Russia, but the impact on the overall budget decreased. Possible reasons include, first, the complete readjustment of the definition of GNP; second, the relatively stronger decrease in civilian nongovernmental net value added compared to the decrease in central government expenditures; and third, the impact of hyperinflation. Furthermore, the most important economic variables for the Russian economy are decreasing—not only military production, but also civilian production declined. The decrease in overall industrial production reached 18.8 percent for 1992, 16.2 percent for 1993, and 24 percent for 1994.[34] The economic conversion losses could therefore not be compensated through civilian production.

Figure 11.1 Military Expenditure Indicators for the USSR/Russia

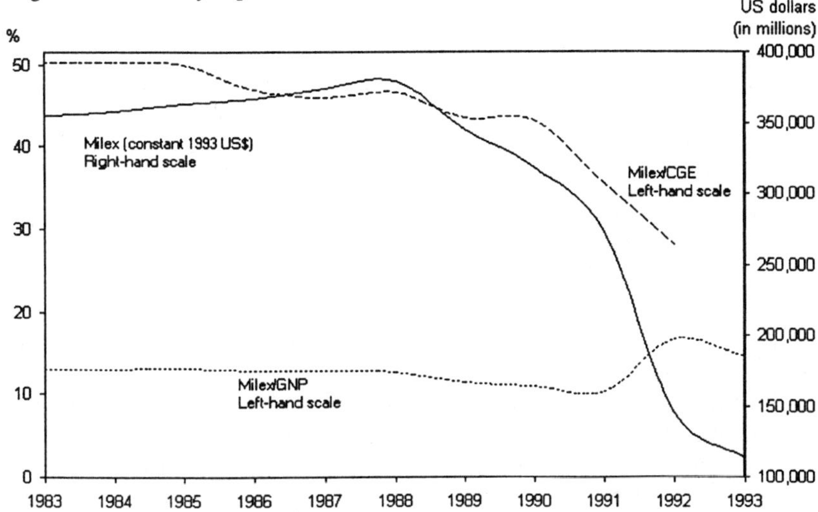

Source: US Arms Control and Disarmament Agency (ACDA), *World Military Expenditures and Arms Transfers 1993–1994* (Washington, D.C.: U.S. Government Printing Office, 1995).

Figure 11.2 Military Expenditure Indicators for China

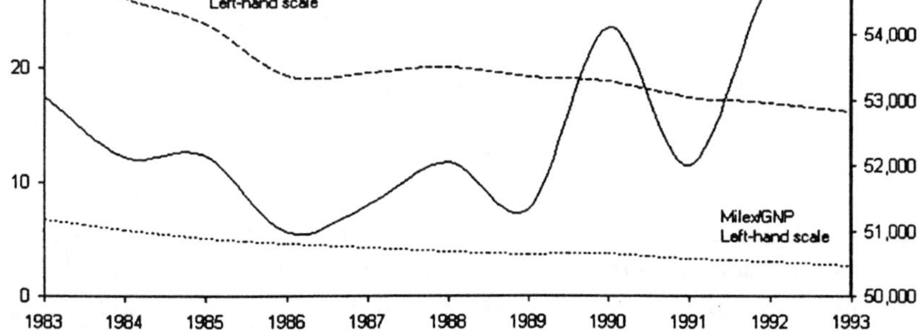

Source: US Arms Control and Disarmament Agency (ACDA), *World Military Expenditures and Arms Transfers 1993–1994* (Washington, D.C.: U.S. Government Printing Office, 1995).

In real terms, military expenditure increased in the USSR until 1988; since then massive reductions have occurred. In China, some shifts can be found, but on a lower level as a whole. The relatively high level in 1983 was reduced, with a few fluctuations, until 1989. As a result of the Tiananmen Square crackdown and, later, the effects of the Gulf War—during which China learned about the effectiveness of modern warfare—the military received bigger outlays after 1989.

Even if resource conversion took place in China as well, the absolute figures indicate that the amount of money spent on the military has been increasing. In other words, the Chinese threat potential has gone up in spite of resource conversion, precluding such qualitative conversion as occurred in Russia. In addition, the economic boom has made it easier for the Chinese to fund defense costs.[35]

From the allocative point of view, the reduction of the defense burden should have generated a peace dividend.[36] However, the two countries do not have the same room for maneuver in transforming the reduction into concrete achievements: China has much more latitude than does Russia. Russia is limited on the one hand by a continuing high dependency on armaments and on the other hand by the missing growth in nondefense business. The constant change of policies has also played a role by preventing strategies from being completely implemented and thus lessening the positive effects of conversion measures due to contravening side effects. Moreover, one must ask whether maneuverability is enough to solve the conversion problems that continue to exist. Analyzing the relationship of the following factors to conversion may help answer this question.

Location. In both countries, military production centers are not distributed uniformly. In Russia, approximately 35 percent of arms-producing companies are located in the Ural, Central, and Northwest regions. A striking concentration is also found in Moscow and St. Petersburg as well as the Moscow and Leningrad oblasts, which together account for approximately 18.3 percent of the total industrial output of the defense complex.[37]

In China, a distinction can be made between Third Line regions and other regions. About 55 percent of all defense factories are located in Third Line regions.[38] Within the Third Line there are specific concentrations along the Yangtze River (Yichang, Wanxian, and Chongqing), but most defense industries are scattered within the large territory consisting of such provinces as Shaanxi, Sichuan, Yunnan, Guizhou, and Hunan and the western part of Hubei.[39] In cases in which concentrations are themselves centralized in smaller geographic entities and the corresponding industrial sector loses ground in terms of relative importance, examples taken from industrial development show that concentrations and monostructures then generate barriers that retard change.[40]

The question of whether armament concentrations impede conversion cannot, however, be answered in a definitive way. On the one hand, regional proximity between companies can result in common strategies (e.g., with regard to the announcement of mass layoffs) based on the concentration of interests, placing them in a position to halt conversion procedures. On the other hand, common action presupposes similar intentions, and similar intentions can arise only if the relevant perspectives do not vary greatly.

The success of conversion as observed through location is difficult to assess in Russia because military and civilian production have been decreasing in most regions. Individual studies even came to the conclusion that the Novosibirsk region, which is as dependent as the St. Petersburg region in terms of armaments but is by far more remote, shows a more positive production development pace. This can be explained by several factors: (1) less confrontation with civilian imported goods, (2) the acceptance of higher indebtedness to other enterprises (i.e., a greater tendency to send goods even if payment is uncertain), and (3) the presence of more conversion support from the state (specifically, for the nuclear industry).[41] This example cannot be generalized, however, because the stated reasons might disappear as economic liberalization continues. In the future, imported goods will reach the more remote regions, the acceptance of higher indebtedness to other companies will be limited, and the will to invest more money in conversion projects will depend on structure rather than region.

The lesson that may be drawn from the numerous case studies now available is that arms manufacturing concentration is not the only indicator of regional impact.[42] Although the arms industry has many particularities (monopsonistic market structure, production like modern custom tailoring, very long production stages, very specialized investment linked to high irreversibilities, production in an isolated environment, etc.), it has never been a monostructural sector, especially in Russia. The share of civilian goods manufactured in the arms sector reached about 40 percent in regional armament centers such as Nizhni Novgorod or about 30 percent in Udmurtia.[43] A close observation of production reveals that the goods were very diversified and came from nearly every civilian sector. Therefore, it seems that concerted action is more likely to take place in regions distant from the center, which are characterized by isolation and remoteness from suppliers and outlets.

In China, conversion has been particularly successful in the special economic zones and in other coastal regions.[44] One should take into account that most production centers have been newly established or restructured with conversion funds. However, there are conversion examples in several big cities that have been successful without the creation of new production centers.[45] Such success stories are, as a percentage, much more infrequent in Third Line locations. In these regions, most military companies are located in areas with little infrastructure and are therefore isolated.[46] The Chinese government, conscious that the location drawbacks would prevent commercial success, launched a program aimed at improving the location conditions (through faster connections to the supplier and sales market) by shifting entire production centers (or parts of them) to more favorable regions.[47] These shifts unfortunately lacked a strategic plan, with which possible sectoral "inherent potential" could have been reached. In some sectors concentrations have been achieved, especially in the vehicle construction industry (vans, cars, motorbikes) in Chongqing and in the optical industry in Wuhan, and the shift to a better location in itself has been an improvement for most companies.[48] An assessment of the achievements of Third Line policy demonstrates that it is the correct policy from the point of view of regional economy.

Nevertheless, two things are missing for a more efficient enforcement of the relocation program: first, sectoral concentration in more cases, which could have led to a focusing of existing potential; and second, institutional restructuring. It would have been possible to separate the companies to be converted from the state sector or at least to restructure them according to product sectors. For television set production alone—one of the pet sectors of conversion—manufacturers are subordinate to the Ministry for Energy Resources, the Ministry of Machine Building, and the Ministry of Aeronautics and Aerospace Industries. In this structure, the subordination of manufacturers to a ministry based on function is relatively unimportant, because of the new purpose of the ministries—which have become more like huge holding companies.

Labor. As a rule, well-developed arms industries are characterized by highly qualified, well-trained personnel as well as higher remuneration than for comparable positions in the civilian sector. Moreover, working in the military sector is often prestigious, especially in sectors with state-of-the-art manufacturing.

These criteria—good qualifications, better wages (in socialist states often linked to indirect privileges such as better access to scarce consumer goods, lodging, etc.), and prestige—fully applied in the case of Russia, in which military production was carried out at advanced world standards. However, the reduction in defense expenditures and the launching of conversion led to a shift in these criteria. Wages are no longer the highest in the country; they are often paid months late, if at all.[49] Uncertainties regarding the meaning of production—to put it clearly, against what enemy—led to a large decline in prestige. However, the need for high qualifications and good training levels remain.

For China, the assessment of the labor factor depends greatly on the type of production and on the sector (e.g., nuclear sector, missiles) and partly on the location. The arms industry is much less developed than the Russian one; many sectors still lag decades behind the world level. In spite of this, China's arms personnel are also characterized by high qualifications and work in a privileged sector.[50]

In both countries, conversion efforts to date have shown that the arms industry is no longer special. There are now alternatives: The civilian sector contributes increasingly more new jobs requiring high qualifications, offering better remuneration, and, in the end, characterized by a certain amount of prestige. For employees, the problem can be reduced to two crucial requirements: First, they must be paid for their work, and second, their work must be appropriate. The analysis of both countries has shown that the arms sector alone can no longer ensure this. Can conversion, with the shift to civilian production, provide sufficient assistance? Are the qualifications still required?

The civilian market is an alternative in both economies. Compared to the Organization for Economic Cooperation and Development (OECD) countries, however, both have far to go in terms of development. Experiences in East Asia indicate that there exists a shortage of labor in economies that must catch up and are in a phase of growth and export orientation. This observation applies especially to qualified personnel, which are a requirement of the arms industry.

Great differences in terms of growth emerge from the comparison between the two countries. China's economy has been growing in the double digits (also due to export promotion policies) for several years, whereas Russia's has been declining on terms quite unfavorable for industrial production. In addition, regional growth is not distributed uniformly across both countries. In China, the coastal regions drive growth, whereas the Third Line sector—in which most arms companies are found—is in constant decline in relative terms. In Russia, disparities are also widening.[51]

With few exceptions, conversion enterprises have thus far not been able to drive economic growth and to develop their potential, even in locations with less competition.[52] Meanwhile, great waves of layoffs are unheard-of in China; the "three irons" (secure jobs, position, and wages) are supposed to be broken up, but only very gradually. Although companies in the state-run sector can now become bankrupt, the ones that do so are certainly not the important arms enterprises—production for national security must be maintained—but rather those authorized to convert due to their insufficient production structure. As for Russia, large reductions have been recorded in absolute terms, but the decline in employment lags severely behind the drop in production levels.[53]

Thus far, both countries' strategies to address the social problems that accompany conversion involve appeasement more than confrontation. On the one hand, this enables the government to cut costs and redistribute the proceeds to cover social benefits. On the other hand, it is a major handicap for individual companies as excessive staff further reduces profitability, so that the position of these companies will probably worsen due to growing domestic competition. A few job programs aimed at retraining and improving qualifications have been launched. Most of these programs occur in firms with foreign involvement, mainly in joint ventures.[54] Such models are thoroughly positive, but they can help only if the companies succeed in producing marketable goods.

Even if conversion remains the alternative, the lack of enforcement measures or of positive results could result in redundancies in both countries in the long run. Although the current growth of the Chinese economy may be better able to absorb this qualified surplus workforce than is the case in Russia, the job opportunities will not necessarily occur in the regions in which the arms production centers are now located. At the same time, workers' movement to regions with jobs is not necessarily desired, given the large migration problems that already exist in China. A small portion of the surplus workforce will have the opportunity to move, but certainly not the majority.

In Russia, large workforce migrations have not been experienced to date—labor remains relatively immobile. This lack of flexibility can probably be explained less by the failure to appreciate opportunities and the attachment to native regions than by the scarcity of accommodations in the developing centers of St. Petersburg and Moscow, the prices of which outweigh the possible advantages for most workers. In addition to general social problems, which, if they lead to unemployment, may result in domestic political problems, Russia is confronted with the problems of restraining atomic mercenaries and transferring know-how, the consequences of which could be even farther reaching.[55]

Financial aspects. A financial analysis of the conversion efforts undertaken in the two countries shows that, although conversion has been widespread, economic success was limited to very few companies. Analyses at the company level point to undercapitalization in nearly all cases.[56] At the same time, the high indebtedness of companies to each other is a major problem in both countries. Unfortunately, the lack of capital in a company narrows the margin for new investments, which are prerequisites for successful conversion.

There are three possible ways to cope with this fundamental obstacle. First, at the company level, outsource some activities to win capital for the core business or to promote various forms of participation (offering employee stock options, involvement of domestic and foreign investors). Second, involve the government by requesting subsidies, low interest rates on loans, guarantees, and so on. The third way directly uses the money market.

Both countries make use of all three possibilities, although figures allowing an in-depth comparison between the two are insufficient or unavailable. Russian companies in particular have been trying all forms of company-level capitalization. The problem with such participation—which first and foremost served the political objective of ownership change—is that, in the absence of a financially strong investor, mostly short-term solutions are implemented. For participation models involving personnel, such short-lived projects merely consumed the savings of the participants.[57] Some companies also backed arms exports; where this method succeeded, the virtually inevitable consequence was to discourage conversion—the capital gain was rarely invested in conversion projects.[58] Finally, outsourcing—that is, dismantling the conglomerates by creating individual companies—supplied nearly no capital.

Less information is available for China on this subject. Company-level capitalization is almost nonexistent in China. Given the financial problems defense conversion companies face, direct state subsidies do not allow for new investments into promising economic fields by all companies. The trend in China clearly points in the direction of the capital market.[59]

However, both countries share the fact that the state wishes to cut its involvement or has already done so. The stages described earlier have shown that subsidies have been very low and that even agreed-upon ones have not been paid, as in Russia. Due to government inactivity on the conversion issue, companies are increasingly forced toward the bank sector to cover losses and meet payroll requirements. Loans can help meet these requirements or finance new investment in other production sectors. The fundamental problem with such loans is, who bears the costs when such credit—which was, as a rule, granted with interest rates below the capital market level—cannot be repaid by the company?

Experiences in both countries demonstrate that state-owned companies swing from one financing method to another, trying to service the interest. Such refinancing is only possible with state guarantees (or, in the case of China, with state decisions to refinance—a fully liberalized bank system does not exist). The extension of guarantees or new loans to conversion enterprises represents a significant burden whose relief has been postponed to better times.[60] In addition, the extension of this method to further sectors of state-owned companies not only mortgages the future but also presents a direct danger to the banking sector. This in turn would weaken the exchange rates and substantially affect the balance of payments, with all the associated risks for each economy as a whole. Considering both available data and the overall size of converting companies, the burden is much heavier on Russia than on China.

Institutional Factors and Institutional Will

As mentioned throughout the preceding discussion, the success of conversion in both countries still remains a challenge. Neither economy has been fully successful in converting, and the isolated success stories are by far outweighed by problems and barriers. With regard to stability, there are alarming signs that mass layoffs could occur, with compounded effects in regions with rather unfavorable environments. The involvement of the banking sector in both countries also indicates difficulty in the future. This handicap is larger for Russia, where the absolute figures expressing the defense burden and the lack of growth tend to intensify the effects. With its tremendous growth rates, China is in a better position to cope with these problems.

Nevertheless, the analysis of these economic factors provides only an incomplete reflection of the situation and does not take into account the institutional questions. Decisions on the amount of armaments procured—or, conversely, on the amount of industrial capacity allowed to be converted lie with the government in the form of its representative, the MIC, and, more precisely, in the hands of the MIC top management.

According to the organigrams established in official statements, the highest Chinese authority that issues directives to the MIC is the Central Military Commission (CMC). However, trend-setting decisions are made not only by the CMC but also by influential, retired Party veterans under the auspices of Deng Xiaoping as well as other members of the politburo. This finding alone speaks volumes about the reliability of organigrams structured according to Western understanding. The informal decisions brought about by relations with relatives and, in a more general way, the existence of *guanxi* (informal networks) usually cannot be inserted into a structured system.[61] In many cases, these decisions are also not taken with absolute clarity, so that the individual levels of enforcement—such as the three leading groups of the PLA (General Staff Department, General Political Department, General Logistics Department), or COSTIND—can still play a significant role in daily management.

An absolutely homogeneous MIC reflecting consistent action is conceivable in no country of the world. Resources are always limited, and scarce resources always imply competition to obtain a greater share of the total. If we consider China's defense budget and the wishes of the individual army branches, it quickly becomes obvious that the respective demands outweigh the available resources.[62]

The current analysis has shown that China strives to implement conversion in its broadest sense. China wants to open most of its arms companies to the civilian market, and many efforts have already been made in this direction. Only 10–20 percent of all industries within the defense sector are excluded from this conversion policy.[63]

Nevertheless, the top priority of the Chinese government is improved defense management; in the process, conversion is given room for maneuver as some resources become obsolete. The wish of the government extends further, however, and is somewhat contradictory. On the one hand, conversion should be available as an instrument to ensure the civilian development of the country in some high-tech sectors. On the other, a portion of the profits should be returned to the military industry and reconversion should be possible.

This pragmatic approach toward conversion involves many dangers, for both the world and China itself. Foreign engagement implies better accessibility to advanced technology, in the form of technology transfer to converting companies. Because most of these companies are involved in both the civilian and military sectors, the possibility that this new technology could fuel reconversion cannot be excluded.[64] A kind of "know-how plus" may also occur in China. If this knowledge—generated primarily with the help and technical expertise of the foreign partner—is also useful for military purposes, it will be difficult to keep it entirely within the civilian part of the MIC. This kind of knowledge transfer in particular renders all export control regimes ineffective.

An internal danger may occur if the tendency gives birth to a new self-supporting entity, a "state within the state." In China, a wide range of civilian goods are now manufactured by companies connected in some way to the MIC. The MIC, in turn, has its own transportation facilities (e.g., airlines, ships) and its own material (airports, ports) as well as institutional (military law) infrastructure. In other words, the MIC is no longer totally dependent on government allocations—it can also finance itself.

At the same time, the size and development of new, nonmilitary business sectors give rise to problems of control and possibly also to overreaching. How much conversion, especially involving the PLA, can be accepted without putting the actual task (military readiness) at risk? Or can military readiness no longer be maintained without financial help from the conversion business? A dilemma has thus arisen. Conclusions drawn from this jungle of inadequate information suggest that overreaching is taking place: hence the "Ten Nos" that limit the civilian activities of some troops as well as some kinds of activities.[65] Nonetheless, opposition to these restrictions shows that current dependencies (individuals who receive supplementary income or who have the opportunity to work in civilian outlets) have already reached such a scope that control has become very difficult to ensure.[66] Money generates greed, and the army is no exception to this rule. Greed can also be a positive incentive for increases in value and therefore growth. In the case of China, however, not only could the actual task of the army be carried out inadequately but also the military infrastructure could help create an influential economic sector that increasingly crowds out other sectors.

In Russia, the risk of an interest conflict between civilian and military activities in the MIC is much less acute. The dimension of the actual military sector was overstrained—the defense burden was too heavy to be financed after the reforms and the new assessment of the economy. Conversion was therefore virtually an imperative and remains one; transformation will not occur without conversion. The power of the military lobby has decreased tremendously in recent years; nearly no new procurement is taking place, and the task is now limited to managing the shrinkage.[67] The whole budget to be distributed has also diminished and, especially in the mid-1990s, the agrarian and energy lobbies have gained ground and play an increasing role in the allocation of the budget.

Nevertheless, this does not mean that the institutional environment is conversion friendly. Experience has shown that acting in a market environment is not possible without undergoing a learning process. The dismantling of the state-owned conglomerates to create independent corporations, which is important for a successful conversion, has been hindered for a long time. The inadequacy of the government's support for conversion has also been a handicap, especially because authorities generally provided no incentives. In addition, the perception of conversion in the military sector plays a major role; many companies had to recognize that military skills cannot be easily turned into civilian ones and that production in many sectors does not flourish in a complex, high-performance, and therefore cost-intensive environment. They also had to acknowledge the necessity of producing below their skill levels and the fact that even such production does not automatically lead to economic success. The mentality of shortage economies, according to which produced goods will also be sold, had to be given up.[68] The liberalization of the market quickly demonstrated that the market governs, not the company.

The framework for armaments and conversion has been improved by the division into the three categories (treasury enterprises, state-owned companies, and converting factories) mentioned above. A line has been drawn—at least for the present—between the items that are to be kept in the armaments sector and the greater portion available for conversion.

Conclusion

This comparison has shown that both Russia and China wish to implement conversion, although their understandings of conversion do not correspond on all points. The starting position for conversion in Russia was by far less propitious than in China. Due to the high economic burden of military expenditure, a resource shift toward civilian expenditure was critical for Russia. For China, however, the absence of a conversion policy would have been merely a handicap. The economic environment also points to another comparatively greater drawback for Russia: While the Chinese economy is witnessing incredible growth rates, the Russian economy is shrinking; the decline of industrial production in particular gives cause for concern.

In both countries, conversion is embedded in a comprehensive strategy and is part of systematic transformation. The difference is that the qualitative and quantitative aspects of conversion are by far more distinct in Russia than in China. Russia is undergoing practical demilitarization; such a clear statement cannot be made for China. Conversion the Chinese way is a pragmatic tool used to support general economic development as well as military restructuring. In this context, the microeconomic objective of conversion is to restructure not only arms companies located in remote areas but also the less important and less effective arms manufacturers. At the macroeconomic level, conversion provides a means of acquiring supplementary financing for the military sector. A comparison of conversion phases reveals that the intention is not really to demilitarize—authorities accept demilitarization in the short term to allow strengthening in the long run through tightening and increased efficiency.

The comparison has also shown that conversion does not work well in the two countries. Most companies operate in the red, despite some isolated success stories. Although China probably fares better due to some successful companies—mostly operating under the auspices of the PLA—the overall outcome has been disillusioning for both countries. The military sector is in numerous ways so different from the civilian sector that conversion cannot be managed in the short term. Furthermore, the reorientation of state-owned companies in particular—a problem that has not yet been solved in either country—will not contribute to easing the situation.

In Russia, one of the achievements of conversion is that a large breach has been made in the wall of secrecy. Russia has become more transparent and therefore easier to appraise. Knowledge of the strengths and weaknesses has been improving. This does not apply to China; although conversion has resulted in better information in specific sectors, the Chinese military sector as a whole did not become more transparent. Analyses of China's defense sector must still rely on guesswork to a great extent. The result of inadequate appraisals was experienced with the Soviet Union in the past: The Soviet potential, as a whole, was largely overestimated, and this overestimation was certainly not a negligible variable in the arms race.

The unsatisfactory functioning of conversion gives cause for concern in two areas. First, to what extent can instability contribute to a failure of conversion? Second, to what extent can conversion itself lead to instability? The first question is more acute for Russia if the economic situation does not improve or if the transformation results in nearly unbridgeable disparities. The Russian case bears the risk of increasing conflicts of interest, as the budget continues to shrink. The military lobby could revive itself by grabbing power and improving its share of the total. A bigger share for the military would mean less conversion. If, in turn, conversion is restricted, the burden of military spending will remain heavy and will, as a result, put the whole transformation process at risk.

As for China, the other tendency is the primary risk: Conversion could lead to instability. The result ultimately depends on the power and the influence of the "state within the state" that is currently emerging. To what extent does the success of the civilian branch of the PLA in particular depend on military power and backscratching and to what extent can it undermine it? Will the remaining sectors be crowded out?

The last question one must ask is whether or not this kind of MIC will be able to maintain a defense industrial base in the future. Will the "state within the state" continue to be controlled? (Controlled, of course, both by the Chinese leaders and by the military sector itself.) If so, to what extent? Even for the military, control is becoming more difficult. Such initiatives as the Ten Nos aimed at forbidding or limiting some conversion business are suffering strong opposition. These reactions reveal the already established intricacy of mutual dependencies.

Notes

The author would like to thank John Frankenstein (Hong Kong University), Joseph DiChiaro III, and Alisa Federico (both of the Bonn International Center for Conversion) for interesting discussions and valuable comments.

1. Calculations based on constant 1984 U.S. dollars. See U.S. Arms Control and Disarmament Agency (ACDA), *World Military Expenditures and Arms Transfers 1987* (Washington, D.C.: U.S. Government Printing Office, March 1988). As for the argument that the Soviet Union would have armed itself regardless of costs, the figures reveal that the USSR was not able to keep pace from the beginning of the new arms race. Upon examining these figures, one is not surprised that the Soviet Empire disintegrated in the end, but rather that few saw it coming.

2. For the historical context see June Teufel Dreyer, "The Role of the Armed Forces in Contemporary China," in *The Armed Forces in Contemporary Asian Societies,* eds. Edward Olsen and Stephen Jurika Jr. (Boulder, Colo.: Westview Press, 1986), pp. 25–54; Paul Humes Folta, *From Swords to Plowshares? Defense Industry Reform in the PRC* (Boulder, Colo.: Westview Press, 1992), pp. 11ff.; Yitzhak Shichor, "Military-to-Civilian Conversion in China: From the 1980s to the 1990s," Working Paper No. 142 (Canberra: Australian National University Research School of Pacific Studies, December 1993); John Frankenstein, "The People's Republic of China: Arms Production, Industrial Strategy and Problems of History," in *Arms Industry Limited,* ed. Herbert Wulf (Oxford: Oxford University Press/SIPRI, 1993), pp. 271–319.

3. See the speech "Generalsekretär des ZK der KPdSU und Vorsitzender des Präsidiums des Obersten Sowjets der UdSSR, Michael Gorbatschow, vor der Generalversammlung der Vereinten Nationen am 7. Dezember 1988," in *Europa-Archiv,* Series 1 (1989), p. D34f.

4. The Russian glasnost (general openness and transparency) does not describe exactly what China's dissident Wei Jingsheng presented in December 1978 in addition to the other four official modernizations. The "fifth modernization" signifies "democratization," which nevertheless cannot be implemented without wider openness and more transparency.

5. In contrast, the United States reached about 6.4 percent at the peak of the arms race in 1987. For further sources, see Hans-Hermann Höhmann and Christian Meier, "Die Wirtschaft Osteuropas im Umbruch," *Beiträge zur Konfliktforschung* 4 (1990), pp. 35–57; Alexei Izyumov, Leonid Kosals, and Rozalina Ryvkina, "The Russian Military-Industrial Complex: The Shock of Independence," in *Conversion of the Defense Industry in Russia and Eastern Europe,* Report 3, ed. Joseph DiChiaro III (Bonn: Bonn International Center for Conversion, April 1995), pp. 12–18.

6. For details, see Johann Adolf Graf Kielmansegg and Oskar Weggel, *Unbesiegbar? China als Militärmacht* (Stuttgart/Herford: Seewald Verlag, 1985), pp. 150ff.

7. In the United States, the ratio in the year 1988 was considerably lower at 13.4; my calculation is based on ACDA data for 1988.

8. The thesis according to which the Soviet Union was "arming to death" and therefore would have been forced to convert overlooks the fact that the "acceptance of suffering" of the population in dictatorships is much higher than in democracies (the imperturbable North Korean way is a good example). The Soviet Union had not nearly reached the share of military expenditure to GNP that was realized during World War II. Pressure did come from the expensive arms race, but the real constraint came from glasnost, which had already been launched.

9. For a comparison of army strength and weaponry for the year 1978, see International Institute for Strategic Studies, *Military Balance 1978–1979* (London: Brassey's Defence Publishing, 1979).

10. Jin Zhude and Chai Benliang, "Strategic Thinking of China's Conversion in the 1990s," in *Proceedings of the International Conference on International Cooperation in Peaceful Use of Military Industrial Technology,* ed. Chai Benliang and Zhang Junping (Beijing: International Academic Publishers, 1992), p. 28; and Folta, *From Swords to Plowshares?* pp. 49ff.

11. See Huai Guomo and Yu Zonglin, "China's Defense Conversion Policy, Its Development," in *China's Defense Conversion,* ed. Cao Shixin, in *Foreign Broadcast Information Service, Daily Report China,* 27 July 1995, p. 9.

12. Ibid., p. 9; Folta, *From Swords to Plowshares?* pp. 63ff.

13. The "16-Character Slogan"; see Chapter 1 (Box 1.1) in this book.

14. See Huai and Yu, "China's Defense Conversion," p. 10.

15. For the historical development and its impact, see John Wilson Lewis and Xue Litai, *China's Strategic Seapower: The Politics of Force Modernization in the Nuclear Age* (Stanford: Stanford University Press, 1994), pp. 214ff.

16. An extensive documentation, also on "window" outlets, can be found in Cao Shixin and Jin Zhude, *Zhongguo Guofang Gongye Junmin Jiehe Shi Zhounian Zhuanji* (Beijing: Bingji Gongye Chubanshe, n.d.); Jin Zhude, ed., *Guide to International Corporation and Investment with Enterprises of China's Defense Industry,* 2 vols. (Beijing: China Association for Peaceful Use of Military Industrial Technology, 1993).

17. Data from China Defence Science and Technology Information Centre, Beijing.

18. See also Wendy Frieman, "China's Defence Industries," *Pacific Review* 6:1 (January 1993), p. 58.

19. For the development and its embedding, see Folta, *From Swords to Plowshares?* pp. 65ff.

20. See Huai and Yu, "China's Defense Conversion," p. 11.

21. COSTIND was created in 1982 for the promotion of specialization, interindustry cooperation, and standardization in the defense business.

22. CAPUMIT was actually founded in the second half of 1985, but official documents indicate that it began operation in 1986.

23. Huai and Yu, "China's Defense Conversion," p. 11. In addition, about 9 billion yuan had been allocated by the provinces. See Jean-Claude Berthélemy and Saadet Deger, *Conversion of Military Industries to Civilian Production in China: Prospects, Problems and Policies* (OECD Development Center Report, final draft, July 1995), p. 44.

24. See Jin Zhude, "Share-Holding System: A New Attempt at Conversion from Military to Civilian Industry in China," in *Restructuring the Military Industry: Conversion for the Development of the Civilian Economy,* eds. China Association for

Peaceful Use of Military Industrial Technology and United Nations Department of Development Support and Management Services (Hong Kong: Publishing House of Electronic Industry, 1994), pp. 94–100; see also Chapter 4 in this book.

25. See *Zhongguo Junzhuanmin Bao* [China Defense Conversion News], 2 June 1992, p. 1, and 2 April 1993, p. 1.

26. See Chen Chunmei, "Defense Serves Dual Purpose," *China Daily,* 14 April 1995, p. 1.

27. At the United Nations Development Programme and CAPUMIT's jointly sponsored conferences in Chongqing, Yichang, and Wuhan in November-December 1995, more than two hundred companies from Third Line regions presented their products and company profiles to the international market, some of them for the first time.

28. Based on a conversation with a military attaché based in Hong Kong, November 1995.

29. Berthélemy and Deger, *Conversion,* p. 53, and private conversations in Chongqing, November 1995.

30. This section is based mainly on Bonn International Center for Conversion (cited hereafter as BICC), ed., *Conversion Survey 1996: Global Disarmament, Demilitarization and Demobilization* (Oxford: Oxford University Press, 1996), p. 130f.

31. See Kevin P. O'Prey, *A Farewell to Arms: Russia's Struggles with Defense Conversion* (New York: Twentieth Century Fund Press, 1995), pp. 75ff; and BICC, *Conversion Survey 1996,* p. 132.

32. A monopolistic position could be maintained until 1991 in some sectors; for other consumer goods, such as washing machines, bicycles, motorbikes, and vacuum cleaners, the arms industry accounted for more than 70 percent of total Russian production. See Petra Opitz, *Krisenmanagement in der russischen Rüstungsindustrie,* Report 6 (Bonn: BICC, October 1995), p. 40.

33. The data do not necessarily correspond with the respective official national figures. For instance, official Chinese defense statistics do not take into account research and development, modernization of defense industry plants and equipment, and various compensation plans. ACDA data are about three times higher than data from official Chinese sources. The current data on Russia are improving; after the dissolution of the Soviet bloc, it was discovered that some ACDA data had not been high enough. As mentioned earlier, the data should be viewed as indicators of trends. For an interesting observation on data, different definitions, and intentions, see Håkan Wiberg, "Measuring Military Expenditures: Purposes, Methods, Sources," *Cooperation and Conflict* 18:3 (September 1983), pp. 161–177.

34. See Deutsches Institut für Wirtschaftsforschungs, Berlin, Institut für Weltwirtschaft an der Universität Kiel, and Institut für Wirtschaftsforschung Halle, "Die wirtschaftliche Lage Rußlands: Fortsetzung des Niedergangs ohne hinreichenden Strukturwandel," *DIW Wochenbericht,* no. 47–48 (1994), p. 806.

35. The example of Japan should be mentioned here. After World War II, Japan was allowed to spend no more than 1 percent of GNP on military expenditures. Due to an economic boom, however, Japan's rank as one of the world's biggest defense spenders jumped in twenty years from position 11 to 5 in 1993 according to ACDA data.

36. For an economic approach and a theoretical discussion of the concept of peace dividend, see BICC, *Conversion Survey 1996,* pp. 63ff.

37. See Ksenia Gonchar, "Conversion of the Arms Industry," unpublished paper commissioned for BICC, August, 1995, p. 18. For an overview of regional defense concentrations in 1985 in the former Soviet Union, see Julian Cooper, "Defense Industry Conversion in the East: The Relevance of Western Experience,"

in *External Economic Relations of the Central and East European Countries (Colloquium 1992),* ed. Reiner Weichhardt (Brussels: NATO, 1992), pp. 165f.

38. See Bates Gill, "Defensive Industry," *Far Eastern Economic Review,* 30 November 1995, p. 62.

39. Approximately one-fourth of China's entire population lives in these regions.

40. For example, in the German Ruhr Valley, a region highly dependent on coal industries and steel mills, some areas were never sold due to higher than expected competition for employees when new businesses were initiated. In addition to these institutional inflexibilities in dealing with a changing environment, other barriers evolve due to high contamination and other costs for reuse as well as a poor image problem.

41. See Opitz, *Krisenmanagement,* p. 29.

42. See also Julian Cooper, *The Soviet Defense Industry: Conversion and Economic Reform* (New York: Council on Foreign Relations Press, 1991); Alexander Ozhegov, "Conversion and Russia's Regions," in *The Post-Soviet Military-Industrial Complex,* ed. Lars B. Wallin (Stockholm: The Swedish National Defense Research Establishment, 1994), pp. 49–63; Nicola A. Mögel, Thomas Sachse, and Hans-Henning Schröder, *Chancen und Probleme der Rüstungskonversion in der Gemeinschaft Unabhängiger Staaten: Konversionsprofile ausgewählter Regionen—Nizhnij Novgorod, Republik Udmurtien, Ekaterinenburg, Republik Belarus,* Report 2 (Bonn: BICC, March 1995).

43. Mögel, Sachse, and Schröder, *Chancen und Probleme,* pp. 13 and 20.

44. See Mel Gurtov, "Swords into Market Shares: China's Conversion of Military Industry to Civilian Production," *China Quarterly* (June 1993), pp. 230ff.

45. For example, the Jiangnan Shipyard in Shanghai and Beijing North Vehicle Manufactory.

46. The openness policy could not oppose this trend. Some of the preferences that had been in effect for a long time in coastal regions were not granted to the southwestern regions before the Ninth Five-Year Plan (1996–2000).

47. If the whole company was not relocated, some cases suggest that civilian parts were shifted first. In the period of the Seventh Five-Year Plan (1986–1990), according to an interview in Wuhan in December 1995, about three-fourths of the funds were used for relocation and reorganization in the Third Line. Many defense producers had to be relocated anyway, because they resided in the region that will be flooded after the Three Gorges Dam Project is completed.

48. Wuhan is not a Third Line region, but in terms of location it is affected by Third Line policies because many defense companies from the Third Line have been relocated to the Wuhan area.

49. See, for example, Dong Clarke, "Far Eastern Military Shipyard Workers Get Some Back Pay," *OMRI Daily Digest,* no. 143 (25 July 1995).

50. This involves more than material privileges. Due to the shortage economy, the arms industry was always a priority. Therefore, production possibilities had greater chances to be used than in the civilian sector.

51. For instance, the region of Yekaterinburg, the city of Zelenograd, and the Russian part of the Donets Basin—just to name a few—are not diversified enough and depend on sectors that are in economic decline.

52. Jianshe motorbikes in Chongqing might be one exception for China, and Shtamp, a machine builder in Tula, for Russia. For the Russian example and an analysis of Russian defense companies, see Antonio Sánchez-Andrés, "The Transformation of the Russian Defence Industry," *Europe-Asia Studies* 47:8 (1995), pp. 1269–1292.

53. See Ksenia Gonchar, "Employment Aspects of Defense Conversion in

Russia," in *Conversion of the Defense Industry in Russia and Eastern Europe,* Report 3, ed. Joseph DiChiaro III (Bonn: BICC, April 1995), p. 57.

54. For an overview of some firms and their experiences, see Andrei Baev, Matthew J. von Bencke, David Bernstein, Jeffrey Lehrer, and Elaine Naugle, "American Ventures in Russia," report of a workshop held on 20–21 March 1995 at Stanford University (Stanford: Center for International Security and Arms Control, May 1995).

55. See Andreas Heinrich and Heiko Pleines, "Russia's 'Nuclear Flea Market' Tempts Smugglers," *Transition,* 17 November 1995, pp. 9–11.

56. See David Bernstein, ed., *Defense Industry Restructuring in Russia: Case Studies and Analysis* (Stanford: Center for International Security and Arms Control, Stanford University, December 1994); Berthélemy and Deger, *Conversion,* p. 61ff.

57. The first phase of privatization in Russia consisted mainly of offering free vouchers that could be used to buy shares of the enterprise in which the voucher holder worked or to trade at auctions. Nonetheless, the purpose was merely the change of ownership and, with few exceptions, did not generate the needed capital for new investments. For a discussion of the overall privatization process in Russia, see Alexander Bim, Derek Jones, and Thomas Weisskopf, "Privatization in the Former Soviet Union and the New Russia," in *Privatization in Central and Eastern Europe,* ed. Saul Estrin (New York: Longman Publishing, 1994), pp. 252–278.

58. See Alexei Arbatav, "Economy and Security in Exports of Combat Aircraft," in *The Arms Production Dilemma,* CSIA Studies in International Security, no. 7, ed. Randall Forsberg (Cambridge, Mass.: The MIT Press, 1994), p. 103.

59. Nevertheless, the debt overhang in both countries, especially for unprofitable firms, is a substantial danger. In addition—as indicated on many factory visits in the last three years—obtaining the "appropriate" amount of loans is becoming a severe problem for some firms.

60. In addition, most companies are weighed down with accumulated debts, partly because of the inability of one enterprise to pay another for goods and services. This "debt chain" constitutes another burden postponed to "better times."

61. One example is the "princelings" (*taizi*) problem, or the endowment of children and relatives of high Chinese cadres with "better" positions and "better" access to power in the "new," very market-oriented Chinese business world.

62. See Bates Gill and Taeho Kim, *China's Arms Acquisitions from Abroad: A Quest for Superb and Secret Weapons,* SIPRI Research Report No. 11 (Oxford: Oxford University Press, 1995), pp. 98ff.

63. According to Ding Henggao (minister and party committee secretary to COSTIND), this applies only to army companies. See Daniel Kwan, "Officials Deny Raising Funds for Arms Buildup," *South China Morning Post,* 6 July 1993, p. 10. Huai Guomo, then deputy director of COSTIND, stated that for the overall defense sector, about 10–20 percent are excluded from producing for conversion. See Huai Guomo, "Open Up New Areas of Conversion Through Enhanced International Cooperation," unpublished paper for the Conference on Conversion—Opportunities for Development and Environment, Dortmund, 24–27 February 1992, p. 6.

64. Alarming signals arise in some high-tech joint ventures, such as the Chinese-American Hua Mei (a telecom company) on a deal with advanced communication technologies, and the dual-use machines deal between McDonnell Douglas and China National Aero-Technology Import and Export Corporation (CATIC). In the latter, it was clearly stated that the items should be used only in a plant near Beijing and only for civilian purposes. The plant was never built, however; instead, parts of the machines ended up at Nanchang Aircraft Manufacturing Corporation, which produces motorcycles, jet-trainers, and crop-spraying planes as well as A-5

strike aircraft and Silkworm missiles. See Bruce Gilley, "Peace Dividend," *Far Eastern Economic Review,* 11 January 1996, p. 1416; Nigel Holloway, "Playing for Keeps," *Far Eastern Economic Review,* 8 February 1996, pp. 14–16.

65. The so-called Ten Nos are regulations prohibiting certain business activities by the military. See also Chapters 9 and 10 in this book.

66. One should note the embezzlement in many of these companies, in the sense that they "failed" to indicate the extent of real profits.

67. For references to the general downsizing, see Elisabeth Sköns and Ksenia Gonchar, "Arms Production," *SIPRI Yearbook 1995: Armaments, Disarmament and International Security* (Oxford: Oxford University Press/SIPRI, 1995), pp. 472ff. On army aviation and the lack of funds for new equipment, see Doug Clarke, "General Warns That the Army Aviation Is Facing Extinction," *OMRI Daily Digest* 25 (5 February 1996).

68. This way of thinking is quite familiar, especially in the armaments sector, which was also involved in the formulation of the production needs for armament goods.

12

Conversion at a Crossroads

Jörn Brömmelhörster & John Frankenstein

Our long-term and continuing studies of defense industry conversion around the world led us very early to the conclusion that a full examination and understanding of the process requires a multifaceted approach. The process—as we hope has been revealed in these studies of conversion in China—is complex and requires more than the usual public policy perspective. In the Chinese case, we must also understand the contexts—political, social, economic—of defense conversion. We must remember that defense conversion involves business propositions, and we therefore must consider commercial and managerial perspectives as well. Finally, due to the special position of the Chinese People's Liberation Army—as political arbiter and a quasi–independent kingdom—relative to the rest of Chinese society, we must deal with military issues and the intriguing question of the entrepreneurial army.

We are also confronted—indeed, stimulated—by what we see as misperceptions of the Chinese experience in defense industry conversion. Misunderstanding of the entire Chinese reform process is not uncommon: Witness the rosy scenarios peddled by twentieth-century descendants of the Enlightenment enthusiasts who saw in China an ideal state and whose passions gave rise to the baroque fantasy of chinoiserie.[1] Today, however, the talk is of a capitalist China, a market of 1.2 billion people, a superpower led by miraculous mandarins—thirty years ago, of course, the Cultural Revolution stimulated comparable foreign passions.

Naturally, we do not deny China's substantial post-1979 progress and even more considerable promise. In fact, it would not be honest to disregard China's accomplishments in building the defense sector that it currently possesses; for instance, Yitzhak Shichor demonstrated in Chapter 6 how the aviation industry went from maintenance through copy-production to attempts at independent design, and in Chapter 4 Chai Benliang pointed out that over the past forty years China has built a comprehensive defense industrial system.

But China is embarked on the messy, contradictory, and uncertain path of modernization, and it is unrealistic to expect linear progress.[2] It is that context that must be kept in mind when we come to the issue of defense industry conversion—it is embedded in the entire construct of the Chinese reform process.

Thus, there are common threads—the paradoxes and conflicts among perception, strategic intent, and conversion results—that run through the chapters in this book. To be sure, an edited collection by its very nature lacks the kind of consistency we might expect in other publications, particularly when the intention is to apply varying disciplinary perspectives to the same subject. Nevertheless, our contributors have all dealt with similar themes and motifs. If conversion, defined by China as civilian production rather than as a shift of military assets to civilian use, is presented to the world as a success, the intent—expressed repeatedly by China's leaders—is to contribute economically and technologically to defense modernization. The economic results are problematic, however, and the strategy, which has its own internal contradictions, appears to emphasize local benefit: the preservation of defense assets and jobs. In sum, the long-term goal remains a strong and modern China.

These issues were laid out first by John Frankenstein (Chapter 1), who, drawing on a range of written materials plus personal observation, provided an overview of the strategic setting and dilemmas of Chinese defense industry conversion. Kay Möller delved further into the international and military contexts in presenting the paradox that defense conversion as a policy came at a time in which China's international position was undergoing significant change and new military challenges and responsibilities were being thrust upon China's armed forces—not the time, perhaps, to cut back or convert. Thus, Möller concluded in Chapter 2, the new emphasis on a high-tech armed force capable of projecting offshore power would require a modernized defense sector. In this light, conversion helps to maintain, rather than reduce, the military-industrial complex and provides a consolidation strategy for defense industrial modernization.

Defense conversion is thus an important issue for the survival of the Chinese defense-industrial complex, and it is not surprising to find that the industry as a whole wants to project an image of success. After all, it is always good politics to report what the boss wants to hear, and in a system in which the correct political line is crucial—and the tradition of reporting good results is entrenched—we should not expect anything less.[3] Furthermore, in the attempt to attract foreign investment to this beleaguered sector, not only is success highlighted but the impression is given that conversion is part of military retrenchment (often at international conferences ostensibly providing sober discussions of conversion but usually making pitches for joint ventures).

In fact, as the chapters in this book demonstrate, conversion facilitates China's military modernization drive by keeping military production lines warm, diversifying production capacity, using profits from civilian production to modernize the military, and, importantly, acquiring dual-use technology (both end-use and industrial/process-use). Indeed, taking the case of the aviation industry, Yitzhak Shichor suggested that the aviation industry's considerable diversification—a shift occurred from 95 percent defense-related output at the end of Mao's era to 25 percent only twenty years later—has barely cut into military capabilities. Rather, the military expands by going international: exports, foreign purchases, and, in the case of relations with Russia, offset production.[4]

Nonetheless, in practice the results may not be those intended. At the operational level, the real issue—as with most of the state sector—is to preserve jobs for state workers. In Chapter 8, Richard Latham revealed that China consolidates its military sector by absorping loss-making enterprises (thereby retaining jobs) rather than by rationalizing or "shrinking smart." Profits remain low, much of the sector's civilian production does not meet market requirements, and management structures in this protected industrial sector are slow to change. At the same time, economic pressures on the increasingly independent managers in the defense sector—and in the People's Liberation Army—have given rise to high-level fears that military responsibilities may be neglected. After almost twenty years of conversion policy and changes in the structure—if not practice—of the industry, Chinese officials deeply involved in the effort still raise the possibility that unless the process deepens, the defense industry, now underutilized, will once again become a serious drag on China's overall economic progress.

Arthur Ding's study of the Third Line (Chapter 7) indicated how contradictory these pressures can be. Development in the Third Line regions remains low. China's policy of developing coastal regions first, at the expense of the interior, has negatively affected the Third Line. Indeed, as the Third Line industries are encouraged to open "window operations," their capital and workers have moved to more attractive coastal operations (many in the special economic zones around Hong Kong and Macao).

Ding suggested that the capital still available in the Third Line areas cannot meet the needs of the enterprises left behind in the hinterland. As a result, they may be forced to turn from the original Maoist strategy of self-reliance to dependence on foreign investment. (And Richard Latham, from his perspective as an executive of a major multinational in China, showed just what it will take for a defense sector joint venture to work if it involves foreign investment.)

Chai Benliang, a distinguished Chinese scholar, brought a management systems perspective to the issue of defense industry conversion. He dealt with the pluses and minuses of the Chinese enterprise responsibility system, the new experimental joint stock companies and larger industrial corporations being formed in Chinese industry. He argued that these reforms broadened investment channels and led to quantitative and qualitative improvements in production. At the same time, dependence on the state remains, and none of the organizational reforms can deal satisfactorily with the question of debt that plagues the state sector. He concluded, however, that because producing civilian goods is a more attractive economic proposition than making weapons, maintaining an adequate defense industrial base under tight budget constraints has become problematic.

In Chapter 5, Ka Po Ng examined the defense conversion process through the lens of the specialized Chinese defense conversion press, a forum for both defense industry policymakers and defense industry managers. He followed press trends that ran from encouraging the conservation of defense assets to promoting economic construction. He concluded that the Chinese authorities regard conversion as a double-edged sword that can cut not only from military to civilian but also from civilian to military production. Conversion, then, is seen by China as a process that will promote both economic efficiency and military capability. But given negative trends—declines in quality, a brain drain, the unwillingness of local authorities to accept jurisdiction over formerly centrally controlled defense plants—we actually see a decrease in military production capacity and thus, potentially, military strength.

The issue of competing interests and goals also appeared in discussions of the role of the army in defense conversion. Tai Ming Cheung demonstrated how PLA expansion into the commercial sector is a response to 1980s reductions in defense appropriations and manpower, just as conversion is an industrial response to diminished demand. These cuts rendered unnecessary substantial portions of the army's logistics supply and production capabilities. At the same time, commanders have had to find ways to generate supplementary funds to maintain troops' living standards. These demands, in addition to the opening of the economy, provided new opportunities for the army—opportunities upon which the army has become dependent, according to Min Chen (Chapter 10). Thus, mirroring other expressions of concern about the effects of conversion on equity and military preparedness, the authors in this book have indicated that coastal units are better off than those in the interior and that there are fears that the PLA's entrepreneurial drive will be stronger than its fighting spirit.

But the success or failure of conversion must ultimately turn on economic results and the contribution of the effort to the larger economy. Here, too, the outcome appears mixed. In a macroeconomic analysis, Feng-Cheng Fu and Chi-Keung Li estimated in Chapter 3 that despite the proliferation of products coming from conversion, the overall contribution of the effort to the national economy is only about 2 percent. According to their investigation, the economic performance of converted enterprises is substantially lower than those of the civilian economy; a comparative examination of defense and nondefense plants in the automobile industry indicated that manufacturers from defense are saddled with surplus labor and low productivity. However, their calculations showed that converted enterprises have a higher export ratio than do civilian concerns. They concluded that although conversion has had some success in areas in which special expertise can be brought to bear—automobiles, for instance—in general, problems of excess inventory and quality remain. According to official statistics, 30 percent of converted enterprises still cannot stand on their own, and the actual percentage may be even higher. Essentially, conversion has allowed the state to lessen financial pressure on the defense sector—another angle on the crucial issue of providing jobs.

The gains or shortcomings of conversion can best be assessed when a country's experiences are juxtaposed against those of a country facing similar challenges. When conversion occurs simultaneously with structural economic reform—as is the case in both China and the former Soviet Union—the process is even more difficult. Using a comparative approach, Jörn Brömmelhörster explored this issue and concluded that Russia and China have followed different conversion policies: China's can be described as trial and error in confined areas, whereas the Russian approach can only be termed as "muddling through"—a process that undermines the potential for success. Although both conversion approaches share a rather disenchanting economic outcome, China's has much better prospects. Whereas China's economy can potentially cope without conversion, the success of conversion is crucial to Russia's economic transformation and political stability. Nevertheless, in Brömmelhörster's opinion China's economic success does not protect it from other evils, such as the potential for destabilization arising from a misuse of the defense sector's resources and infrastructure.

Finally, we must ask a paradoxical question: If Chinese defense conversion is, as it appears to be, a frustrated and only partially successful project, if it only commands rhetorical attention from the center and if it contributes only a small amount to China's economy, why should we care? Does it matter?

We think that it does. The problems and paradoxes we have outlined are those of a transitional economy; the reform of China's economy will take a long time, and how reform plays out in this sector will be a measure of the real potential of the economy. China may have brought these problems upon itself, but we should not gainsay its determination to overcome them. Second, and more important, defense industries carry a political weight far greater than their economic contribution. If, as we have argued, conversion is a step on the way to a more modern and efficient defense sector, and if conversion reaches its goals, China's weight in the next century will be great indeed.

Notes

1. The willingness of the West to misunderstand and create myths about China is grist for the mill of many China scholars, from Simon Leys's acerbic writings to the perceptions of Lucian Pye and Jonathan Spence. For a useful wrap-up of this theme, see Spence's essay "Looking East: The Long View," in Jonathan Spence, *Chinese Roundabout: Essays in History and Culture* (New York: W. W. Norton, 1992), pp. 78–90.

2. Samuel Huntington's classic description of the dynamics of modernization provide a remarkable fit to the Chinese situation of the late 1990s. See Samuel Huntington, *Political Order in Changing Societies* (New Haven, Conn.: Yale University Press, 1968), pp. 49–50.

3. For a report of a typical complaint, see "Statistics Cheats Disrupt China's Economic Plans," *South China Morning Post,* 18 August 1994, which begins: "China's macroeconomic planning is being seriously hampered by ambitious officials who falsify statistics to obtain promotion, the official *People's Daily* said in a front page commentary yesterday."

4. For instance, Chinese aircraft exports appear to be predominantly military. According to the *1993 Chinese Industry Yearbook,* between 1979 and 1992 China exported six hundred aircraft, of which "about one-tenth" were civilian planes. See *Zhongguo Gongye Nianjian 1993* [Chinese Industry Yearbook 1993] (Beijing: Zhongguo Gongye Nianjian Chubanshe, 1993), p. 224.

The Contributors

Jörn Brömmelhörster is researcher and project leader at the Bonn International Center for Conversion (BICC), Bonn, Germany.

Chai Benliang is recently retired as professor and senior research analyst at the China Defense Science and Technology Information Center, Beijing, China, and now heads a social science encyclopedia project.

Min Chen is assistant professor at the American Graduate School of International Management, Glendale, Arizona. He is also vice president of Teleflex in Beijing, China.

Tai Ming Cheung is a consultant on China affairs in Hong Kong.

Arthur S. Ding is associate research fellow at the Institute of International Relations (IIR) of National Chengchi University, Taipei, Taiwan.

Feng-Cheng Fu is research fellow at the Chung-Hua Institution for Economic Research, Taipei, Taiwan.

John Frankenstein, formerly at the University of Hong Kong, now is affiliated with the Institute of Politics at the University of Copenhagen and Asian Research Unit at the Copenhagen Business School.

Richard J. Latham is president of United Technologies Corporation, International Operations-China, Beijing, China.

Chi-Keung Li is associate research fellow at the Chung-Hua Institution for Economic Research, Taipei, Taiwan.

Kay Möller is senior researcher at the Stiftung Wissenschaft und Politik (SWP) at Ebenhausen, Germany.

Ka Po Ng is a Ph.D. Candidate at the University of Queensland, Brisbane, Australia, and formerly a lead researcher at the Universities Service Center at the Chinese University of Hong Kong.

Yitzhak Shichor is professor and chair of Chinese Studies at the Hebrew University of Jerusalem, Israel.

Index

About the Book

Mixed Motives, Uncertain Outcomes looks critically at China's efforts to adapt its vast military-industrial complex to the service of its socialist market economy. The authors—all of whom have witnessed or participated firsthand in the country's defense conversion—offer political, macroeconomic, business, and military perspectives on this complex issue.

The book places the conversion process in the context of China's military modernization, the reforms of its business and industrial enterprises, and the international problematic of change in transitional economies. Beijing's underlying strategy, the authors conclude, attempts to combine sometimes conflicting objectives—defense industry modernization at minimal economic and social cost—and its success is by no means assured.

Jörn Brömmelhörster leads the Defense Conversion in the Far East Project at the Bonn International Center for Conversion (BICC). Previously, he was research associate at the Seminar for Economic Policy and Public Finance at the Ruhr University Bochum (Germany) and also managed a group of university lecturers working in the field of defense conversion. **John Frankenstein,** formerly of the University of Hong Kong is currently at the University of Copenhagen and the Copenhagen Business School, teaching courses on China and international management. He is also a fellow at the university's Centre for Asian Studies. His publications include studies for the Stockholm International Peace Research Institute (SIPRI) and the Joint Economic Committee of the U.S. Congress, as well as articles for *China Quarterly,* The Economist Intelligence Unit, *The International Executive,* and *Long Range Planning.*

DURHAM UNIVERSITY LIBRARY
3 0104 01065421 2